MATHEMATICS PROBLEMS WITH SEPARATE PROGRESSIVE SOLUTIONS: HINTS, ALGORITHMS, PROOFS

VOLUME 1: INTERMEDIATE AND COLLEGE ALGEBRA

Cătălin Bărboianu
Evgheni Tokarev

INFAROM Publishing
Scholarly and applied mathematics
office@infarom.com
http://www.infarom.com

ISBN 978-973-88662-9-4

Publisher: **INFAROM**
Authors: **Cătălin Bărboianu, Evgheni Tokarev**
Translator: **Cătălin Bărboianu**
Correction editor: **CarolAnn Johnson**

Copyright © INFAROM 2008

This work is subject to copyright. All rights are reserved, whether the whole work or part of the material is concerned, specifically the rights of translation, reprinting, reuse of formulas and tables, recitation, broadcasting, reproduction on microfilm or in any other way, and storage in data banks.
Duplication of this publication or parts thereof is permitted only under the provisions of Copyright Laws, and permission for use must always be obtained from INFAROM.

CONTENTS

Introduction 5
Problems 11
Hints 27
Algorithms 37
Proofs 67
References 121

INTRODUCTION

Problem-solving is a necessary activity not only for students within the didactic process and for acquiring mathematical proficiency, but also as a primary form of mathematical research and creation.

Proposing and solving problems represent continual scientific training, which the future mathematician begins in classes and continues through his or her individual preparation for school and contests, having as the main goal self-improvement in mathematics, perhaps even pursuing a career in this field.

For problem solvers, this ongoing scholastic training offers the principal way of developing the skills essential to mathematical thinking and technique, as well as achieving some successes in competition. These necessary skills include intuition, selective observation ability, analytical approach, theoretical framing, deduction and logical construction ability, and calculation speed.

However, the didactic material necessary for such training – namely problem books – while abundant in the publication market, for the most part do not reach a professional level which would allow the user to develop those previously enumerated skills.

In considering problem books with solutions, we observe that many of these books remain in the status of general collections of problems and solutions, and are presented as such. Not only do these books not provide solving methodologies, much less systematized methodologies, but also they lack methodic criteria or even affiliation with a certain domain or subdomain of mathematics. Moreover, those problem books do not provide a structure by which to stimulate and enrich individual study so necessary to the mathematical development of students.

In launching this publishing project – a series of problem books in a new structural format – we hope to intervene actively in the processes of exploring, attempting, and effectively solving problems. Our structure moves logically from the approach and theoretical framing of problems, through the steps to be executed, and finally to the generation of a complete solution.

The problem book is structured in four separate and independent sections, namely *Problems*, *Hints*, *Algorithms*, and *Proofs*, in this order.

The *Problems* section consists of 101 problems themselves, which are of medium to advanced difficulty level. These problems were selected from the category of those whose solution is based exclusively on the elementary theoretical results learned in classes.

The solution, however, does not fall into place as a direct application of the theory, but rather, they are the outcome of a non-trivial process of deduction, construction, and observation. These problems are specific to out-of-class mathematics workshops and preparation for mathematics contests.

Each problem has a corresponding position in each of the three sections coming next, which actually contain the progressive solutions to the problem.

The *Hints* offer groups of keywords (which can be words, groups of words, sentences, or short mathematical expressions) that suggest to the solver – intuitively, as well as analytically – an initial approach to the problem, important observations upon which the solution is based, categories of theoretical results applied when solving, and specific theoretical results.

The hints also suggest indirectly the solving algorithm (found in the next section), but without exposing or synthesizing it. All these suggestions, indications, and references are presented in an incomplete, short form, leaving to the solver the task of investigating the various possible approaches and choosing the one that leads to the correct solution.

In the *Algorithms* section, the solving algorithms provide chronological groups of steps necessary for generating the complete solution.

The algorithm is presented as a brief list of tasks; it does not reveal the complete solution to the problem, but only points out the partial tasks whose results will finally yield the logical construction of the solution. These tasks are shown in a short form, without presenting explicitly the mathematical objects upon which to operate, but rather, making precise references to the subjects of the previous steps.

The algorithm indicates directly the correct path for solving the problem, as well as the methodology associated with each step, but does not expose the concrete work, the way of combining the partial results for building the logical constructions, or the detailed development of the calculations.

The reason for this deliberate abbreviation is to allow the solver to discover the exact identification of the objects and working tasks, with the aim of integral reconstitution of the solution.

These features eliminate the risk of a wrong approach to the problem or of following an incorrect or nonproductive path toward a solution; at the same time, they leave enough room for the individual to work toward completing the integral solution.

The *Proofs* represent the complete integral solutions of the problems, unfolded according to the solving algorithm.

This comprehensive presentation includes the detailed steps to be executed, the observations that precede the deductions, and the entire logical motivation. No partial results are left unproved, neither as an exercise nor as being obvious or easily deduced.

Thus the material can be read, followed, and understood by as large a category of problem solvers as possible, not just those with an advanced level of mathematical training.

The sections described above are separated in this book so that the solver can explore the problem and search for solving paths independently, consulting the next section only when he or she has exhausted, with no success, his or her own approach and individual study methods.

As the solver moves progressively from a partial solution to a more complete one, this additional effort itself becomes a useful mathematical exercise. Moreover, the process of moving successively through the indications of the problems together with individual investigation and autocorrection of a wrong approach, stimulates and motivates the solver toward a solution.

All these elements give this type of problem book a truly interactive character.

The problems can be presented and discussed at mathematics workshops and sessions of preparation for contests and olympiads as well as in the classroom, using problems of different difficulty levels with separate groups of students; for advanced groups, an instructor can use this problem book in its entirety.

These problems offer a range of difficulty levels, from non-difficult problems, whose paths and methods of solving can be deduced directly from their wording, to quite difficult problems, and even to a difficulty level of international olympiad problems. Most of the solutions are not immediately obvious and do not represent direct applications of an isolated theoretical result, the theoretic ensemble being necessary for more complex solutions.

Solvers will also encounter problems of the "false difficult" type in which the statement of the problem creates the false impression of a long and laborious solution, while in fact the solution becomes visible as result of observing an important detail or making an ingenious construction or choice. Problems of this type should provide excellent counterexamples for those who tend to label a mathematics problem "impossible" when they fail to discover its solution immediately – and even to attach this label to mathematics itself.

The topics of the problems belong to algebra II and algebra of the first two years of college, passing through fields of integer and real numbers, equations, inequalities, powers, logarithms, divisibility, polynomials, and combinatorics. The problems were selected with care so that each problem meets all the intended methodological criteria, including the difficulty level.

The problem book actually contains three categories of problems, based on their difficulty level. These are arranged consecutively, passing gradually from one level to the next, so that the whole block of proposed problems is a homogeneous one.

The material, by its structure, is useful not only for problem solvers, but also for mathematics instructors, offering a didactic tool that can facilitate the development of students' intuition in approaching medium and advanced level problems, as well as perfecting their algorithmic solving skills.

The problem-solving assumes theoretical and analytical skills, as well as algorithmic skills, coupled with a basic mathematical intuition. The concept of such a problem book successfully supports the development of these skills of the solver and meanwhile offers mathematics instructors models for teaching problem-solving as an integral part of the mathematics learning process.

The present work is the first of a series that will also operate in other domains and subdomains of mathematics. The problem books which follow will be presented in the same structural format, with separate progressive solutions.

This series is part of a far-reaching publishing project whose goal is the involvement of mathematics instructors and graduates in editing such interactive problem books, in similar or different formats, thereby enriching the supplementary didactic material so necessary to study and improvement in contest and scholarly mathematics.

PROBLEMS

AL1.1.1 Solve the system for x, y real:
$$(x-1)(y^2+6) = y(x^2+1)$$
$$(y-1)(x^2+6) = x(y^2+1)$$

AL1.1.2 Consider the sequence of positive integers that satisfies $a_n = a_{n-1}^2 + a_{n-2}^2 + a_{n-3}^2$ for all $n \geq 3$. Prove that if $a_k = 1997$, then $k \leq 3$.

AL1.1.3 Find the least natural number a for which the equation
$$\cos^2 \pi(a-x) - 2\cos \pi(a-x) + \cos\frac{3\pi x}{2a}\cos\left(\frac{\pi x}{2a}+\frac{\pi}{3}\right) + 2 = 0$$
has a real root.

AL1.1.4 Find all natural numbers a, b, c such that the roots of the equations
$$x^2 - 2ax + b = 0$$
$$x^2 - 2bx + c = 0$$
$$x^2 - 2cx + a = 0$$
are natural numbers.

AL1.1.5 Prove that the equation
$x^2 + y^2 + z^2 + 3(x+y+z) + 5 = 0$ has no solutions in rational numbers.

AL1.1.6 Given that $133^5 + 110^5 + 84^5 + 27^5 = k^5$, with k an integer, find k.

AL1.1.7 Prove that $\dfrac{1}{1999} < \dfrac{1}{2}\cdot\dfrac{3}{4}\cdot\ldots\cdot\dfrac{1997}{1998} < \dfrac{1}{44}$.

AL1.1.8 For each natural number $n \geq 2$, determine the largest possible value of the expression
$$V_n = \sin x_1 \cos x_2 + \sin x_2 \cos x_3 + \ldots + \sin x_n \cos x_1,$$
where $x_1, x_2, \ldots, x_n$ are arbitrary real numbers.

AL1.1.9 Determine all primes p for which the system
$$p + 1 = 2x^2$$
$$p^2 + 1 = 2y^2$$
has a solution in integers x, y.

AL1.1.10 Find all real solutions of the system of equations
$$x^3 = 2y - 1$$
$$y^3 = 2z - 1$$
$$z^3 = 2x - 1$$

AL1.1.11 Define the functions
$$f(x) = x^5 + 5x^4 + 5x^3 + 5x^2 + 1$$
$$g(x) = x^5 + 5x^4 + 3x^3 - 5x^2 - 1$$
Find all prime numbers p for which there exists a natural number $0 \leq x < p$, such that both $f(x)$ and $g(x)$ are divisible by p, and for each such p, find all such x.

AL1.1.12 Let $f : (0, \infty) \to R$ be a function such that
(a) f is strictly increasing;
(b) $f(x) > -1/x$ for all $x > 0$;
(c) $f(x) f(f(x) + 1/x) = 1$ for all $x > 0$.
Find $f(1)$.

AL1.1.13 Find all integer solutions of $\dfrac{13}{x^2} + \dfrac{1996}{y^2} = \dfrac{z}{1997}$.

AL1.1.14 Given five consecutive positive integers whose sum is a cube and such that the sum of the middle three is a square, find the smallest possible middle integer.

AL1.1.15 Let $f(n) = (1^2+1)1! + (2^2+1)2! + \cdots + (n^2+1)n!$. Find a simple general formula for $f(n)$.

AL1.1.16 Show that for every odd integer n the sum $1^n + 2^n + \cdots + n^n$ is divisible by n^2.

AL1.1.17 For any positive integer k let $f_1(k)$ denote the sum of the squares of the digits of k (when written in decimal), and for $n \geq 2$ define $f_n(k)$ iteratively by $f_1(f_n(k))$. Find $f_{2007}(2006)$.

AL1.1.18 Let $a_1 < a_2 < \cdots < a_{43} < a_{44}$ be positive integers not exceeding 125. Prove that among the 43 differences $d_i = a_{i+1} - a_i$ ($i = 1, 2, \ldots, 43$) some value must occur at least 10 times.

AL1.1.19 Let $a, b, c > 1$ be real numbers, and let $S = \log_a bc + \log_b ca + \log_c ab$. Find the smallest possible value of S.

AL1.1.20 Let $\overline{abc}$ represent a three digit number in base 10, with $a \geq c + 2$. Let $\overline{efg} = \overline{abc} - \overline{cba}$. Show that $\overline{efg} + \overline{gfe}$ is constant, for all a, b, c, as above.

AL1.1.21 Each point in the plane is colored either red or black. Prove that one of these colors contains, for each positive value of d, a pair of points at distance d.

AL1.1.22 A sequence $(a_n)_{n \geq 0}$ of real numbers is defined recursively by $a_0 = 1$, $a_{n+1} = \dfrac{a_n}{1 + na_n}$ ($n = 0, 1, 2, \ldots$).
Find a general formula for a_n.

AL1.1.23 Let $f(x)$ be a polynomial of degree n such that $f(k) = k/(k+1)$ for $k = 0, 1, \ldots, n$. Find $f(n+1)$.

AL1.1.24 Let a sequence (x_n) be given by $x_1 = 1$ and $x_{n+1} = x_n^2 + x_n$ for $n \geq 1$. Let $y_n = 1/(1+x_n)$ and let $S_n = \sum_{k=1}^{n} y_k$ and $P_n = \prod_{k=1}^{n} y_k$ denote, respectively, the sum and the product of the first n terms of the sequence (y_n). Show that $P_n + S_n$ is constant for all n.

AL1.1.25 Suppose that $a_1, a_2, ..., a_n$ are n given integers. Prove that there exist integers r and s with $0 \leq r < s \leq n$ such that $a_{r+1} + a_{r+2} + \cdots + a_s$ is divisible by n.

AL1.1.26 Find the integer part of the real number $a = \log_2 3 + \log_3 5 + \log_5 8$.

AL1.1.27 Given a positive integer n, let n_1 be the sum of digits (in decimal) of n, n_2 the sum of digits of n_1, n_3 the sum of digits of n_2, etc. The sequence (n_i) eventually becomes constant, and equal to a single digit number. Call this number $f(n)$. For example, $f(1999) = 1$ since for $n = 1999$, $n_1 = 28$, $n_2 = 10$, $n_3 = n_4 = \cdots = 1$. How many positive integers $n \leq 2001$ are there for which $f(n) = 9$?

AL1.1.28 Let x, y, and z be non-zero real numbers satisfying $\frac{1}{x} + \frac{1}{y} + \frac{1}{z} = \frac{1}{x+y+z}$. Show that $x^n + y^n + z^n = (x+y+z)^n$ for any odd integer n.

AL1.1.29 Let $N = 9 + 99 + 999 + \cdots + \overbrace{99...9}^{99}$. Determine the sum of digits of N.

AL1.1.30 Suppose a, b and c are integers such that the equation $ax^2 + bx + c = 0$ has a rational solution. Prove that at least one of the integers a, b and c must be even.

AL1.1.31 Define a sequence (a_n) by $a_0 = 0$, $a_1 = 1$, $a_2 = 2$, and $a_n = a_{n-1} + a_{n-2} - a_{n-3} + 1$ for $n \geq 3$. Find, with proof, a_{2004}.

AL1.1.32 For which positive integers n does the equation $a_1 a_2 + a_2 a_3 + \cdots + a_{n-1} a_n + a_n a_1 = 0$ have a solution in integers $a_i = \pm 1$? Explain.

AL1.1.33 Show that expression $\sum_{k=0}^{\left[\frac{n}{2}\right]} (-1)^k C_n^{2k} \cos^{n-2k} \frac{1}{n} \sin^{2k} \frac{1}{n}$ does not depend on n.
C. Bărboianu

AL1.1.34 Find a non-zero polynomial $P(x, y)$ such that $P([a], [2a]) = 0$ for all real numbers a.

AL1.1.35 Show that every positive integer is a sum of one or more numbers of the form $2^r 3^s$, where r and s are non-negative integers and no summand divides another.

AL1.1.36 Find all positive integers $n, k_1, \ldots, k_n$ such that $k_1 + \cdots + k_n = 5n - 4$ and $\dfrac{1}{k_1} + \cdots + \dfrac{1}{k_n} = 1$.

AL1.1.37 Let a, b, c be positive real numbers such that $abc = 1$. Prove that $\dfrac{1}{a^3(b+c)} + \dfrac{1}{b^3(c+a)} + \dfrac{1}{c^3(a+b)} \geq \dfrac{3}{2}$.

AL1.1.38 In a finite sequence of real numbers the sum of any seven successive terms is negative and the sum of any eleven successive terms is positive. Determine the maximum number of terms in the sequence.

AL1.1.39 Solve in positive integers the equation
$$x^y = \left(x + \frac{1-x}{1+y}\right)^{y+1}$$
C. Bărboianu

AL1.1.40 Solve the equation $C_x^y = x + y$ if x and y are positive integers with $x \geq y$.
S. Haufmann

AL1.1.41 A three-digit integer N is the square of integer n. If we switch the two last digits of N we obtain the square of $n + 1$. Find n.
A. Odobescu

AL1.1.42 Show that no pair of natural numbers m and n exists such that expressions $m^2 + 4n$ and $n^2 + 4m$ are both perfect squares.
I. Cucurezeanu

AL1.1.43 Consider the sequence characterized by the recurrence relation $x_n = px_{n-1} + qx_{n-2}$ (x_0 and x_1 are positive, while p and q are integers with $p \geq 1, q \geq 0$). Prove that:
a) $x_n \geq x_{n-1}(p-1) + qx_0 + (p+q)x_1$.
b) If $x_0 = 1$ and $x_1 = p - 1$, expression $E = (p+q)x_n + qx_{n-1}$ is divisible by $p + q - 1$.
D. Ralescu

AL1.1.44 Find number N having the following properties:
$N = \overline{xyztuv}$; $2N = \overline{ztuvxy}$; $3N = \overline{yztuvx}$
$4N = \overline{uvxyzt}$; $5N = \overline{vxyztu}$; $6N = \overline{tuvxyz}$
L. Țene

AL1.1.45 Let M be a finite sequence consisting of n integers (not necessarily distinct). Show that M has at least one non-empty subsequence such that the sum of its elements is divisible by n.
Gh. Szöllösy

AL1.1.46 Given the relations $a \geq c$, $b \geq c$, $ab > 0$, and $(a+b)c = 2ab$, show that $a = b = c$.
C. Ionescu-Țiu

AL1.1.47 Let $f : E \to E$ be a function, where E has a finite number of elements, such that $(f \circ f)(x) = x$, for any $x \in E$. Prove that if E has an odd number of elements, then there exists at least one element $k \in E$ such that $f(k) = k$.
Gh. Ionescu

AL1.1.48 Determine strictly positive numbers $a_1, a_2, \ldots, a_n, \ldots$ satisfying the relation $a_1^3 + a_2^3 + \cdots + a_n^3 = (a_1 + a_2 + \cdots + a_n)^2$ for any $n \in N^*$.
L. Panaitopol

AL1.1.49 Show that number $2^n + 3^n$ is not a perfect square.

AL1.1.50 Let $a + \dfrac{1}{a}$ be an integer number, with a real. Show that number $a^n + \dfrac{1}{a^n}$ is also integer, where n is an integer.
A. Tuțescu

AL1.1.51 Show that the equation $x^n + y^n = z^n$, where n is an integer larger than 1, has no integer solutions x, y and z with $0 < x \leq n$ and $0 < y \leq n$.

AL1.1.52 Let (a_n) be an arithmetic progression of positive integers for which the common difference is prime. Given that the sequence includes both a term that is a perfect j-th power and a term that is a perfect k-th power, and that j and k are relatively prime, prove that there exists a term that is a perfect jk-th power.
S. Amrahov

AL1.1.53 Fix primes p and q. Prove that there are at most six integers x such that the area of the triangle with side-lengths p, q, and x is a positive integer.
K. Boklan

AL1.1.54 Let a and b be positive integers with $ab > 1$. Show that if a and b are relatively prime, then $(1+ab) \nmid (a^2+b^2)$.
K. Holing

AL1.1.55 Consider $m, n \in \mathbb{N}$ with $m+n$ odd. Prove that there is no $A \subseteq \mathbb{N}$ such that for all $x, y \in \mathbb{N}$, if $|x-y| = m$ then $x \in A$ or $y \in A$, and if $|x-y| = n$ then $x \notin A$ or $y \notin A$.
D. Skordev

AL1.1.56 Let a_1, a_2, a_3, a_4 be the consecutive side lengths of a convex quadrangle and let $p, q \in \mathbb{R}_+$ such that $p > \sqrt{a_1^2 + a_2^2}$ and $q > \sqrt{a_3^2 + a_4^2}$. Show that p and q cannot be simultaneously the lengths of that quadrangle's diagonals.
C. Bărboianu

AL1.1.57 Evaluate the sum $\sum_{k=n}^{2n} C_k^n 2^{-k}$.

AL1.1.58 Call a set of integers A double-free if it does not contain two elements a and a' with $a' = 2a$. Determine, with proof, the size of the largest double-free subset of the set $\{1, 2, \ldots, 256\}$.

AL1.1.59 Find the smallest integer $n > 11$ for which there is a polynomial of degree n with the following properties:
  (a) $P(k) = k^{11}$ for $k = 1, 2, \ldots, n$;
  (b) $P(0)$ is an integer;
  (c) $P(-1) = 2003$

AL1.1.60 Let $D = \{d_1, d_2, \ldots, d_{10}\}$ be a set of 10 distinct positive integers. Show that any sequence of 2006 integers from D contains a block of one or more consecutive terms whose product is the square of a positive integer.

AL1.1.61 Solve the equation $\sqrt{[x]} = \lfloor\sqrt{x}\rfloor$ in real non-negative numbers. Generalization.
C. Bărboianu

AL1.1.62 Call an integer n *blocky* if $n > 1$ and there is a run of n consecutive integer squares the average of which is a square. (Thus, 25 is blocky because $\left(\sum_{k=0}^{24} k^2\right)/25 = 4900/25 = 14^2$, and 31 is the next blocky integer.)
a) Determine the set B of blocky integers.
b) Given a blocky integer n, give a procedure that determines all integers k that can serve as starting points for required runs of squares.
c) Give a formula in terms of n for the number of k that can serve as starting points of required runs of squares.
S. Marivani

AL1.1.63 Prove that a necessary and sufficient condition for the graph of polynomial $P(x)$ to admit the center of symmetry $C(a, b)$ is that there exists another polynomial $Q(x)$ such that:
$P(x) = b + (x-a)Q\big[(x-a)^2\big]$, for any real x.
M. Țena

AL1.1.64 Determine the set of values of function $f(x) = \dfrac{x^2 + 2}{\sqrt{x^2 + 1}}$ defined on R.
Șt. Kàdàr

AL1.1.65 Find the number of elements of finite set A, such that to exist bijections from $A \times A$ to $\mathcal{P}(A)$. How many such bijections do exist? (we denoted by $\mathcal{P}(A)$ the set of parts of set A).
H. Banea

AL1.1.66 Let P be a polynomial with positive coefficients and let $x_1, x_2, \ldots, x_n$ be n strictly positive numbers. Show that the following inequality holds true:
$$P^2\left(\frac{x_1}{x_2}\right) + P^2\left(\frac{x_2}{x_3}\right) + \cdots + P^2\left(\frac{x_n}{x_1}\right) \geq nP^2(1)$$
M. Bencze

AL1.1.67 Determine all possible values of expression
$E(a) = \sqrt{a^2 + a + 1} - \sqrt{a^2 - a + 1}$, for a real.
S. Rădulescu

AL1.1.68 Solve the equation in reals:
$$\sqrt{x} + \sqrt{y-1} + \sqrt{z-2} = \frac{1}{2}(x+y+z)$$
T. Andreescu

AL1.1.69 Let $m \geq 1, n \geq 1$ be integers such that $\sqrt{7} - \frac{m}{n} > 0$.
Show that $\sqrt{7} - \frac{m}{n} \geq \frac{1}{mn}$.
R. Gologan

AL1.1.70 Let $A = \{a_1, a_2, \ldots, a_n\}$ be a set of real numbers and let $f : A \to A$ be a bijective function such that:
$a_1 < a_2 < \cdots < a_n$ and $a_1 + f(a_1) < a_2 + f(a_2) < \cdots < a_n + f(a_n)$
Show that f is the identity function of set A.
Does this affirmation still holds true if we replace A with the set of integer numbers?
M. Dădârlat

AL1.1.71 Show that if $p_1 + p_2 + p_3 = 3q \pm 1$, where p_1, p_2, p_3 are three consecutive primes, then q is composite.
I. Cucurezeanu

AL1.1.72 Given two similar right-angle triangles, with side-lengths b, c, respectively b', c' and hypotenuses a, respectively a', show that if $aa' = bc' + b'c$, then the triangles are isosceles.
Şt. Tache

AL1.1.73 The sum of digits of natural number x is equal to y and the sum of digits of number y is z. Find x, given that $x + y + z = 60$.

AL1.1.74 Find a natural number n such that $n = 3\left[\sqrt{n}\right] + 1$. Show that there exist only two such numbers.

AL1.1.75 Given a natural number $m \geq 2$, show that function $f : N \to R$ defined by $f(n) = \{m^n \sqrt{2}\}$ is injective (we denoted by $\{a\}$ the fractional part of a).
C. Niţă

AL1.1.76 Let $x_1, x_2, \ldots, x_n$ be n strictly positive numbers.
a) If $\lg x_1 + \lg x_2 + \cdots + \lg x_n < 0$ and $\lg x_1 \lg x_2 \cdots \lg x_n > 0$, then at least two of the given numbers are smaller than 1.
b) If $\lg(x_1 x_2 \cdots x_n) = p$ and $\lg^2 x_1 + \lg^2 x_2 + \cdots + \lg^2 x_n = q$, express as function of p and q the sum
$S_n = \lg x_1 \lg(x_2 x_3 \cdots x_n) + \lg x_2 \lg(x_3 x_4 \cdots x_n) + \cdots + \lg x_{n-1} \lg x_n$.
A. Ghioca

AL1.1.77 Let $x_1, x_2, \ldots, x_n$ be n non-zero numbers. Prove that there exists an irrational number a such that numbers ax_i, $i = 1, \ldots, n$ are all irrational.
N. Ceti, V. Marchidan

AL1.1.78 Show that number
$N = 1 + 2 \cdot 3 + 4 \cdot 5 \cdot 6 + 7 \cdot 8 \cdot 9 \cdot 10 + 11 \cdot 12 \cdot 13 \cdot 14 \cdot 15 + \cdots$ is not perfect square, if the number of its terms is larger than 1.
C. Rusu

AL1.1.79 Find the remainder of number 5^{7^n}, $n \in N$, upon division by 31.
D. Andrica

AL1.1.80 Prove that $|\sin 2x + 2\sin(x+y) - \sin 2y| \leq 2\sqrt{2}$, for all reals x, y.
D. Bătinețu

AL1.1.81 Given 1000 non-zero and distinct natural numbers having the sum 1000998, prove that there are at least two odd numbers among them.
L. Niculescu

AL1.1.82 Solve in natural numbers the equation
$(1 + x!)(1 + y!) = (x + y)!$
T. Andreescu

AL1.1.83 Solve in reals the equations $x \cdot 2^{\frac{1}{x}} + \frac{1}{x} \cdot 2^x = 4$.
L. Panaitopol

AL1.1.84 If n is a natural number divisible by 5, then number $N = \underbrace{11\ldots10}_{n}^2 - n^2$ is divisible by 225.
I. Voicu

AL1.1.85 Let $A = na^2 + pb^2$ and $B = (n+p)(a^2 + b^2)$ be two expressions, where a, b, n and p are integers. Show that if expression A is divisible by product $(p - n)(a - b)$, then expression B is also divisible by the same product.
Şt. Tache

AL1.1.86 Let a, b, c be the side-lengths of a triangle, satisfying: $a - b \geq 0, b - c \geq 0$. Given that the area of the triangle is $2n^2$, show that $a + b > 2n$. If the triangle is right-angled, then $4n^2 < 2a^2 - c^2$.
C. Joița

AL1.1.87 Let $(a_n)_{n\geq 0}$ be a sequence of real numbers such that $a_{n+1} + a_{n-1} = \sqrt{2} \cdot a_n$, $\forall n \geq 1$. Show that the sequence is bounded.
D. Bătinețu

AL1.1.88 Let p be the number of divisors of number $N_1 = n^4 - 4n^3 + 8n^2 - 8n + 5$, $n \in N$ and let q be the number of positive divisors of number $N_2 = N_1 - 1$. Prove that $p + q$ is odd.
A. Ghioca

AL1.1.89 Prove that any integer k can be represented in an infinity of ways in the form $k = \pm 1^2 \pm 2^2 \pm \cdots \pm m^2$, for certain natural numbers m and certain choices of signs plus and minus.
Erdös – Surányi problem

AL1.1.90 Given $b > 1$ and $d > 0$ such that $2^b > b^d$, then
$$\log_b(n+1) > \frac{d}{bn} + \log_b n, \forall n \in N^* \text{ and}$$
$$\log_b(n!) > \frac{dn}{b}\left(\frac{1}{2} + \frac{1}{3} + \cdots + \frac{1}{n}\right), \forall n \geq 2.$$
D. Bătinețu

AL1.1.91 Solve in natural numbers the equation $x(x+2)(x+8) = 3^y$.
T. Andreescu

AL1.1.92 Prove that sequence $(a_n)_{n \in N}$ defined by
$$a_{n+1} = \frac{2}{2 - a_n}, \ a_1 \neq 2, \text{ is periodic.}$$
T. Andreescu

AL1.1.93 Solve in integers the equation $1 + x + x^2 + \cdots + x^{2n} = y^{2n}$, $n \in N^*$.
M. Lalescu, O. Pop

AL1.1.94 Consider the sequences $(u_n)_{n \in N}$ and $(v_n)_{n \in N}$ that satisfy the recurrence relations
$u_{n+1} = 3u_n + 4v_n$, $v_{n+1} = 2u_n + 3v_n$, $u_0 = 3$, $v_0 = 2$.

Let $(x_n)_{n \in N}$ and $(y_n)_{n \in N}$ be the sequences defined by $x_n = u_n + v_n$, $y_n = u_n + 2v_n$. Prove that $\left[\sqrt{2}x_n\right] = y_n$, $\forall n \in N$.

D. Andrica

AL1.1.95 Find all polynomials $P \in R[X]$ such that $P(x^2) = P^2(x)$, $\forall x \in R$.

AL1.1.96 Show that whatever five integers we choose, there are two among them, whose sum or difference is divisible by 7.

I. Tomescu

AL1.1.97 Show that $\pi(n!+2n) + \pi(n) \leq \pi(n!+n) + \pi(2n)$, where $\pi(x)$ is the number of primes less than or equal to x.

L. Panaitopol

AL1.1.98 Let m, n be odd natural numbers and let $P_{n,m}(x) = (x-1)^n(x+1)^m = a_0 + a_1 x + \cdots + a_{n+m}x^{n+m}$ be a polynomial. Denote by $c_{n,m}$ the number of null coefficients a_i, $i \in \{0, 1, 2, \ldots, n+m\}$ of polynomial $P_{n,m}$.

a) Show that $c_{n,m}$ is odd number.

b) Let $n, m \geq 3$. Show that the condition $n = m$ is sufficient for the inequality $c_{n,m} \geq 3$ to hold true. Is this condition also necessary?

M. Țena

AL1.1.99 Let $p \geq 3$ be a prime number. For every $k \in \{1, 2, \ldots, p-1\}$ denote by a_k the remainder of k^p upon division by p^2. Show that $a_1 + a_2 + \cdots + a_{p-1} = \dfrac{p^3 - p^2}{2}$.

AL1.1.100 Prove that there exist an infinity of natural numbers a such that for any natural number n, number $P = n^4 + a$ is not prime.

AL1.1.101 Let P be a polynomial with complex coefficients. Prove that its associated polynomial function is even if and only if there exists a polynomial Q with complex coefficients such that $P(x) = Q(x)Q(-x)$, $\forall x \in C$
M. Țena

HINTS

AL1.1.1 Adding and subtracting the equations; completing squares; substitutions.

AL1.1.2 Reductio ad absurdum; the given relation holds for two different triples of four consecutive terms; inequalities between numbers that differ each other by a positive quantity.

AL1.1.3 Completing a square; sum of non-negative terms is 0; when a cosine is 1 and −1; in integers, if $2 \mid 3x$, then $2 \mid x$.

AL1.1.4 The discriminants are perfect squares; in natural numbers, if $n < p^2$ and n is a perfect square, then $n \leq (p-1)^2$.

AL1.1.5 Sum of squares; substitutions; reductio ad absurdum; every perfect square is congruent to 0 or 1 modulo 4; theorem of division; divisibility; parity.

AL1.1.6 The last digits of n and n^5 are the same; the left-hand side is congruent to 0 modulo 3; inequalities.

AL1.1.7 $\dfrac{n}{n+1} < \dfrac{n+1}{n+2}$; factorials; combinations; $\sum_{k=0}^{n} C_n^k = 2^n$; $C_{1998}^{999} > C_{1998}^k$.

AL1.1.8 $2ab \leq a^2 + b^2$; $\sin^2 x + \cos^{2x} = 1$.

AL1.1.9 Parity; congruence modulo p.

AL1.1.10 $x = y = z$; reductio ad absurdum; inequalities.

AL1.1.11 $f(x) + g(x)$; congruence modulo p.

AL1.1.12 Substitution $x \to f(x) + 1/x$; injectivity.

AL1.1.13 gcd(x, y); square-free numbers; divisibility.

AL1.1.14 If prime p divides k^2, then $p^2 \mid k^2$; divisibility.

AL1.1.15 $n(n+1)!$; induction.

AL1.1.16 Grouping the terms in pairs; development of binomial $(a-b)^n$.

AL1.1.17 Cyclic sequence; period; congruence modulo the period.

AL1.1.18 The sum of the differences; reductio ad absurdum; the minimum of the sum.

AL1.1.19 Properties of logarithm; arithmetic mean – geometric mean inequality.

AL1.1.20 Uniqueness of decimal representation.

AL1.1.21 Reductio ad absurdum; two numbers, one for each color; circle around a black (or red) point; existence of a chord of length smaller that the radius.

AL1.1.22 Substitution; iteration; the sum of the first m natural numbers.

AL1.1.23 Polynomial $(x+1)f(x) - x$; roots; factorization.

AL1.1.24 Relation between x_n and y_n; $\dfrac{1}{1+x} = \dfrac{1}{x} - \dfrac{1}{(1+x)x}$.

AL1.1.25 Partial sums S_n; counting principle; congruence modulo n; $S_s - S_r$.

AL1.1.26 Arithmetic mean – geometric mean inequality; properties of logarithm; inequalities.

AL1.1.27 Criterion for divisibility by 9; multiples of 9.

AL1.1.28 Clearing denominators; common factors; product equals to zero; x^n is an odd function.

AL1.1.29 Powers of 10; digit 1.

AL1.1.30 Reductio ad absurdum; fraction in lowest terms as solution; parity.

AL1.1.31 $b_n = a_n - a_{n-1}$; recurrence for b_n; induction; b_{2n}; a_{2n}.

AL1.1.32 n divisible by 4; sum of odd number of terms each ± 1 cannot equal 0; reductio ad absurdum for $n = 2m$, with m odd; $a_{n+1} = a_1$; summation from 1 to n; product of the terms of that sum.

AL1.1.33 Moivre's formula for $(\cos x + i \sin x)^n$; $x = 1/n$; development of $(a+b)^n$; real part and imaginary part.

AL1.1.34 $[a]+[b] \leq [a+b] \leq [a]+[b]+1$; consecutive integers.

AL1.1.35 Complete induction; even and odd cases; every positive integer lies between two consecutive powers of 3.

AL1.1.36 Arithmetic mean – harmonic mean inequality.

AL1.1.37 Cauchy inequality; arithmetic mean – geometric mean inequality.

AL1.1.38 The sequence of partial sums s_n; inequality chain starting with $0 < s_{11}$ and finishing with 0.

AL1.1.39 The equality in arithmetic mean – geometric mean inequality.

AL1.1.40 $x \geq y+2$; $C_x^y > x+y$ for $y \geq 3$; induction.

AL1.1.41 The decimal notation; $10 \leq n+1 \leq 31$; parity.

AL1.1.42 Reductio ad absurdum; order; if $x > n^2$ and x is a square, then $x \geq (n+1)^2$ in integers; parity.

AL1.1.43 Summing from 2 to n; $\sum x_i$; factorization; positiveness; integer fraction.

AL1.1.44 Number of digits is constant, maximum of x; order between x, y, z, t, u, v; decimal notation; divisibility.

AL1.1.45 Sequence of partial sums; reminders upon division by n; counting principle; $a \equiv b \pmod{n} \Rightarrow a-b \vdots n$.

AL1.1.46 Positiveness; $a \geq b$ and $b \geq a$; harmonic mean.

AL1.1.47 Inverse of the function; bijectivity; reductio ad absurdum; parity.

AL1.1.48 $n = 1$; $a_n = n$; induction; formulas of sums of similar powers of the first n natural numbers.

AL1.1.49 Parity; last digit; 3 does not divide $2^n + 3^n$; $(a \pm b)^n$; multiples of 3; reductio ad absurdum.

AL1.1.50 $a^{-n} = \dfrac{1}{a^n}$; induction.

AL1.1.51 $x = y$; divisibility by 2; order between x and y; decreasing sequence of even positive integers; for integers a and b, if $a > b$, then $a \geq b+1$; formula of $z^n - y^n$.

AL1.1.52 Congruence between the terms of the progression; Fermat's little theorem.

AL1.1.53 Theorem of sines; Pitagora's generalized theorem; inequalities between the side-lengths of a triangle; properties of the primes; transformation of sine and cosine of the same argument into fractions with two integer variables.

AL1.1.54 Theorem of division; reductio ad absurdum; order between a and b; the least a; in integers, if $x\,|\,y$, then $x\,|\,y + zx$; in integers, if $t\,|\,s$ and $s < 2t$, then $t = s$.

AL1.1.55 Reductio ad absurdum; $A \neq \phi$; if $x \in A$, when $x + mn \in A$?; iteration; the sum of two natural numbers having the same parity is even.

AL1.1.56 Ptolomeu's inequality; Cauchy-Buniakowsky-Schwartz inequality; reductio ad absurdum.

AL1.1.57 $n = 1$; expression in n is constant; combinatorial properties; splitting sums.

AL1.1.58 Reductio ad absurdum; the odd part of a; partition of A by the possible odd parts of a; ranges of the odd parts; cardinality.

AL1.1.59 Polynomial $Q(x) = P(x) - x^{11}$; roots; factorial form; divisibility.

AL1.1.60 Product of the first n terms: P_n; P_n / P_m perfect square; P_n is a product of powers of d_i; the parity of the exponents; the parity of the difference; the counting principle.

AL1.1.61 Perfect squares; $p^2 + 1 \leq (p+1)^2$; n-th root; $p^n + 1 \leq (p+1)^n$.

AL1.1.62 $n \equiv \pm 1$ or $n \equiv \pm 7$ (mod 24); difference of sums of squares; formula of sum of the first squares; integers; congruence modulo 6; completing square; factors with same parity whose product is even; congruence modulo 4; factorizations of $(n^2 - 1)/12$; number of positive divisors of $(n^2 - 1)/48$.

AL1.1.63 Relation between $P(a-h)$ and $P(a+h)$; substitution $y = x - a$; in particular, $x = a - h$, $x = a + h$.

AL1.1.64 Variable $\sqrt{x^2+1}$; $a > 0 \Rightarrow a + \dfrac{1}{a} \geq 2$; biquadratic equation.

AL1.1.65 equal cardinalities; $2^n > n^2$ from a certain n upward; induction; $n^2 > 2n + 1$ from a certain n upward.

AL1.1.66 Arithmetic mean – geometric mean inequality for the powers of $\dfrac{x_i}{x_j}$; multiplication by coefficients of P; Cauchy-Buniakovsky-Schwarz inequality for $P\left(\dfrac{x_i}{x_j}\right)$ and terms equal to 1.

AL1.1.67 $E(-a) = -E(a)$; formula of difference between two squares; $E(a) < 1$; equation $E(a) = b$, with $b \in (0, 1)$.

AL1.1.68 Substituting the radicals; completing squares; sum of squares equals to 0.

AL1.1.69 The least k such that $m^2 + k \vdots 7$; $7n^2 - m^2 \geq 3$.

AL1.1.70 Induction; reductio ad absurdum; $p < n$ such that $f(a_p) = a_n$; $n - p + 1$ numbers larger than a_p; $f(x) = x + 1$.

AL1.1.71 Arithmetic mean; parity; consecutiveness.

AL1.1.72 Pitagora's theorem; proportional side-lengths.

AL1.1.73 Number of digits of x; $y \leq 9$ or $9 < y \leq 18$; table of values.

AL1.1.74 $a - 1 < [a] \leq a$; raising to the second power; inequations.

AL1.1.75 Reductio ad absurdum; $\sqrt{2}$ is irrational; $\{a\} = a - [a]$.

AL1.1.76 Properties of logarithm; reductio ad absurdum; product of numbers smaller than 1 is smaller than 1; positiveness; $2s_n$; formula of $\left(\sum_{i=1}^{n} y_i\right)^2$.

AL1.1.77 If m is prime, then $\sqrt{m}$ is irrational; table with n columns and an infinity of rows; reductio ad absurdum; counting principle; a, b rational implies a/b rational.

AL1.1.78 Terms multiples of 10; remainder upon division by 10; last digit.

AL1.1.79 5^{3k+1}; $125 = 124 + 1$.

AL1.1.80 Formula of $(\sin a - \sin b)$; division by $\sqrt{1+\sin^2(x-y)}$; properties of functions sin and cos.

AL1.1.81 Reductio ad absurdum; minimum of the sum of 1000 even numbers; parity.

AL1.1.82 Case $x \geq 2$, $y \geq 2$; parity; divisibility; symmetry of the equation.

AL1.1.83 $x > 0$; arithmetic mean – geometric mean inequality; $\frac{1}{x} + x \geq 2$.

AL1.1.84 $225 = 9 \cdot 25$; $a, b \vdots 5 \Rightarrow a^2 - b^2 \vdots 25$; formula of difference between two squares.

AL1.1.85 $B - 2A$; factorization; formula of difference between two squares; divisibility.

AL1.1.86 The height is smaller than the adjacent sides; $2n < 4n$; Pitagora's theorem; $n^2 < 2n^2$.

AL1.1.87 a_{n+2}; a_{n+4}; a_{n+8}; periodicity.

AL1.1.88 N_2 is a perfect square; the number of divisors of the square of a natural number is odd; $x^2 + 1$ is not a perfect square.

AL1.1.89 $(m+1)^2 - (m+2)^2 - (m+3)^2 + (m+4)^2 = 4, \forall m \in N$; induction; $m \to m+4$.

AL1.1.90 $\left(1+\dfrac{1}{n}\right)^n \geq 2, \forall n \in N^*$; properties of logarithm; sum the first inequality over $k = 1, \ldots, n$; induction; factorization; splitting sum; factor out $\dfrac{d(n+1)}{b}$.

AL1.1.91 Powers of 3; substitutions of the form 3^z; divisibility.

AL1.1.92 a_{n+2}; a_{n+4}.

AL1.1.93 Cases $x > 0$ and $x < -1$; $\sum\limits_{i=0}^{2n} x^i$ is between x^{2n} and $(x+1)^{2n}$; in integers, there is no integer between two consecutive numbers; formula of $a^m - b^m$; binomial theorem.

AL1.1.94 $y_{n+1}^2 - 2x_{n+1}^2 = y_n^2 - 2x_n^2$; $x^2 < x^2 + 1 < (x+1)^2, \forall x > 0$; x_0, y_0.

AL1.1.95 $P = 0$, $P = 1$; writing P in its standard form; $P = a_n X^n$; reductio ad absurdum; if polynomials P and Q are equal and P contains a term in X^m, then Q also contains such a term.

AL1.1.96 Remainders of the perfect squares upon division by 7; theorem of division; counting principle; formula of difference between two squares; divisibility.

AL1.1.97 $A = \{n, n+1, \ldots, 2n\}$, $B = \{n! + n, n! + n + 1, \ldots, n! + 2n\}$, $f : A \to B$ bijective; bijectivity; f transports composite numbers in composite numbers; number of primes from A and B.

AL1.1.98 $P_{n,m}\left(\dfrac{1}{x}\right)$; $a_{n+m-j} = -a_j$; $a_{\frac{n+m}{2}}$; parity; $P_{n,n} = Q(x^2)$; the coefficients of the odd-degree terms are zero; $P_{7,15}$; formula of difference between two squares; binomial theorem; $a_3 = a_{11} = a_{19} = 0$.

AL1.1.99 Theorem of division for k^p and $(p-k)^p$; reductio ad absurdum; divisibility; binomial theorem; in N^*, if $a < 2b$ and $b \mid a$, then $a = b$.

AL1.1.100 Factorization; formula of difference between two squares; $\sqrt{2\sqrt{a}} = 2m \in N$; factors larger than 1.

AL1.1.101 Standard form of P; identification of coefficients; $P(x) = P_1(x^2)$; roots of P_1; canonical form of P_1; formula of difference between two squares.

ALGORITHMS

AL1.1.1
Add the two given equations together.
Complete squares in the resulting equation.
Subtract the second given equation from the first.
Group the terms and factor out to get
$(x-y)(x+y-2xy+7) = 0$.
Rearrange the terms and factor out in the second factor of the product.
Do the substitutions $a = x - 5/2$ and $b = y - 5/2$.
Solve the new system in a and b formed by the last equation and the one obtained after adding together the two given equations.
Arrange the terms such to put in evidence $a + b$.
Solve the quadratic equation in $a + b$.
Do a new subtraction and arrange the terms to put in evidence $a - b$.
Find $a - b$.

AL1.1.2
Reductio ad absurdum. Assume $k > 3$ for which the given relation holds.
Consider the four consecutive terms that must exist.
Apply the given relation for the fourth term $a_k = 1997$ and show that $a_{k-1} \leq 44$.
Apply the given relation for the third term a_{k-1} and show that $a_{k-1} \geq 61$, contradiction.

AL1.1.3
Show that the equation holds if $a = 6$, $x = 8$.
Complete a square and write the equation as a sum of two non-negative terms.
Equal each term to zero.
From the first term being 0, show that $x \equiv a \pmod{2}$.

From the second term being 0, show that $3x \equiv -2a \pmod{12a}$ or $3x \equiv 4a \pmod{12a}$.

Show that a is divisible by 2 and by 3, which proves that the least a is $a = 6$.

AL1.1.4
Put the condition on the discriminants to be perfect squares.
Show that $b \geq 2a - 1$ and the analogue relations.
Add together the three analogue inequalities.
Show that $(a, b, c) = (1, 1, 1)$ is the only possibility.

AL1.1.5
Complete squares and write the left side of the equation as a sum of squares.
Show that the given equation is equivalent with $a^2 + b^2 + c^2 = 7d^2$ in integers.
Reductio ad absurdum: assume a minimal solution (in sense of the sum of the absolute values).
Show that every perfect square is congruent to 0 or 1 modulo 4, using the theorem of division.
Show that a, b, c, d are even.
Show that $(a/2, b/2, c/2, d/2)$ is also a solution, which contradicts the induction hypothesis of having a minimal solution.

AL1.1.6
Show that the last digit of the left-hand side is 4.
Show that $3 \mid k$ and the smallest possibility for k is 144, the next is 174.
Find that each power from the left-hand side is smaller than a multiple of 10^{10} and add them together.
Show that the left-hand side is smaller than 10^{11}.
Show that k cannot be 170, hence it cannot be 174.

AL1.1.7
If p is the product of fractions, denote $q = \dfrac{2}{3} \cdot \dfrac{4}{5} \cdot \ldots \cdot \dfrac{1998}{1999}$ and prove that $p < 1/44$ by considering pq.
Write p as a fraction whose numerator is a factorial product.

Show that $p = 2^{-1998} C_{1998}^{999}$ and use $\sum_{k=0}^{n} C_n^k = 2^n$ to show that $p > 1/1999$.

AL1.1.8

Use the inequality $2ab \le a^2 + b^2$ and show that $V_n \le \dfrac{n}{2}$.

Show that the value $n/2$ is reached for some x_i.

AL1.1.9

Assume $x, y \ge 0$ and show that p is odd.
Show that $x \equiv \pm y \pmod{p}$ and then $x + y = p$.
Show that $p = 4x - 1$, x is 0 or 2 and then $p = 7$.

AL1.1.10

Find the solutions with $x = y = z$.
Show that these are all the solutions; reductio ad absurdum.
Assume $x > y$. Show that $y > z$ and $z > x$.

AL1.1.11

Add together $f(x)$ and $g(x)$ and factor the result.
Show that p divides either 2, x, $x+1$, or $x + 4$.
Study each possibility by using the congruence modulo p.

AL1.1.12

Substitute $k = f(x) + 1/x$.
Apply (c) to k.
Apply again (c) to x and equal the quantities.
Using injectivity of f, show that $f(x) = \dfrac{1 \pm \sqrt{5}}{2x}$.
Show that only the "$-$" option satisfies all three conditions.

AL1.1.13

Denote $d = \gcd(x, y)$ and rewrite the equation in x_1, y_1 (the quotients of the division of x, respectively y, with d).
Show that $x_1^2 \mid 1997 \cdot 13$ and $y_1^2 \mid 1997 \cdot 1996$.

Show that (x_1, y_1) is $(1, 1)$ or $(1, 2)$ and consider them as separate cases.

Replace them back in the equation for each case and find d and then the solutions (x, y).

AL1.1.14
If n is the middle integer, show that $n = 3k^2$ for some k.
Show that k is a multiple of 15 and take the smallest multiple.
Find n.

AL1.1.15
Show that $f(n) = n(n + 1)!$ by induction:
Check it for $n = 1$. For $n \geq 1$, assume the formula holds for n.
Write $f(n + 1)$ as a sum, with $f(n)$ being one of its terms.
Use the assumption for n.
Factorize the sum with the common factor $(n + 1)!$.
Factorize again the second factor and check that the claimed formula holds true for $n + 1$.

AL1.1.16
Check for $n = 1$.
Assume $n > 1$ and group the terms from the sum:
Leave the last term and group the remaining terms symmetrically (the first and the last, etc.).
Write that sum indexed from $k = 1$ to $(n - 1)/2$ and develop its general term as a power of a binomial.
Cancel the similar terms of opposite signs and check that all remaining terms are all divisible by n^2.

AL1.1.17
Start from $k = 6$ and iterate the sum of squares of the digits.
Write the chain and note that the sequence repeats itself, with period 8.
Divide 2007 by 8 and show that $f_{2007}(2006) = f_7(2006)$.

AL1.1.18
Show that the maximum of the sum of d_i's is 124.

Reductio ad absurdum: assume no value occurred more than 9 times; show that the sum is at least 125, contradiction.

AL1.1.19
Write each logarithm of S as a sum of logarithms.
Convert each new term into a 10 base logarithm.
Write S as a sum of six fractions holding 10 base logarithms.
Apply the arithmetic mean – geometric mean inequality to the six-term sum.
Check that the bound is attained.

AL1.1.20
Write all numbers in their decimal representation, by putting in evidence the powers of 10.

Show that $e = a - c - 1$, $f = 9$ and $g = 10 - a + c$, by using the uniqueness of decimal representation.

Compute $\overline{efg} + \overline{gfe}$.

AL1.1.21
Reductio ad absurdum: fix two numbers a and b such that no pair of red points has distance a and no pair of black points has distance b; assume $a \leq b$.

Show the existence of a black point P.

Consider the circle of radius b around P and show that it is colored red.

Show that there exist two points on this circle having distance a, contradiction.

AL1.1.22
Substitute $b_n = 1/a_n$ and find the new recurrence.

Iterate the new recurrence and show that $b_n = b_0 + \sum_{k=0}^{n-1} k$.

Calculate b_n as a function of n.

Come back to a_n and express it.

AL1.1.23
Consider polynomial $g(x) = (x + 1)f(x) - x$ and show that it has $0, 1, \ldots, n$ as roots.
Factorize $g(x)$ with a constant c, putting in evidence these roots.
Set $x = -1$ and find c.
Set $x = n + 1$, find $g(n + 1)$, then $f(n + 1)$.

AL1.1.24
Show that $y_n = x_n / x_{n+1}$ and compute P_n.
Using the identity $\dfrac{1}{1+x_n} = \dfrac{1}{x_n} - \dfrac{1}{(1+x_n)x_n}$ and the given recurrence, compute S_n.
Show that $P_n + S_n = 1$.

AL1.1.25
Let $S_0 = 0$, $S_n = \sum_{i=1}^{n} a_i$. Apply the counting principle to these $n + 1$ integers regarding their remainders upon division by n.
Show that there exists S_s and S_r such that $S_s \equiv S_r \pmod{n}$.
Evaluate $S_s - S_r$.

AL1.1.26
Apply the arithmetic mean – geometric mean inequality for the three logarithms and show that $a > 4$, using properties of the logarithm.
Show that $\log_2 3 < 2$, $\log_3 5 < \dfrac{3}{2}$, $\log_5 8 < \dfrac{3}{2}$.
Show that $a < 5$.

AL1.1.27
Show that $f(n) = 9$ if and only if n is multiple of 9.
Count the multiples of 9 below 2001.

AL1.1.28
Show that $(x + y)(x + z)(y + z) = 0$, by clearing denominators, simplifying, and factoring out in the given relation.

Equal each factor to zero and study the relations between the numbers to the power n in each case.

AL1.1.29
Write each term as a difference between a power of 10 and 1.
Group the powers of 10 and the 1's.
Show that N contains only digits of 1 and 0.
Count the digits of 1.

AL1.1.30
Suppose a, b and c are all odd and $x = p/q$, with $(p, q) = 1$, is a rational solution.

Replace back x in the equation and clear the denominators.

Show that p and q cannot be both even, study all cases and find a contradiction in either case.

AL1.1.31
Let $b_n = a_n - a_{n-1}$. Find a recurrence relating b_n and b_{n-2} by using the recurrence for a_n.

Show by induction that $b_{2n} = b_{2n-1} = n$ for all $n \geq 1$.

Show that $a_{2n} = n(n+1)$.

Calculate a_{2004}.

AL1.1.32
Assume n is divisible by 4 and show that the pattern $(1, 1, -1, -1)$ repeated $n/4$ times is a solution.

Consider the case when n is not divisible by 4. Show that if n is odd, the left-hand side cannot equal 0.

Consider the case when $n = 2m$, where m is odd. Reductio ad absurdum: suppose there exist integers $a_i = \pm 1$ for which the equation holds.

Set $a_{n+1} = a_1$ and write the left-hand side of the equation as a sum from 1 to n.

Show that exactly m of the terms of that sum must be equal to 1 and m must be equal to -1.

Show that the product of all $2m$ terms of that sum is -1.

Show that product also equals to $\prod_{i=1}^{n} a_i^2 = 1$, contradiction.

AL1.1.33

Write Moivre's formula for $(\cos x + i \sin x)^n$.
Write the formula for the particular case $x = 1/n$.
Unfold the binomial of power n in the left-hand side.
Equal the real and imaginary parts respectively in the two sides.
Show that the given expression is equal to $\cos 1$.

AL1.1.34

Show that $2[a] \leq [2a] \leq 2[a] + 1$:
Use $[a] + [b] \leq [a+b] \leq [a] + [b] + 1$ for $a = b$.
Show that either $[2a] = 2[a]$ or $[2a] = 2[a] + 1$.
Write this double sentence as a product that equals to 0.
Replace the values with variables for obtaining the polynomial with the given property.

AL1.1.35

Complete induction: assume every number in $\{1, 2, ..., n-1\}$ can be properly represented. Show the same for n:

If n is even, show that $n/2$ can be properly represented (it is smaller than $n - 1$) and multiply it by 2.

If n is odd, bound n between two consecutive powers of 3 (3^s and 3^{s+1}).

Show that, if $3^s < n$, the number $\dfrac{n - 3^s}{2}$ can be properly represented (it is smaller than $n - 1$) and it is smaller than 3^s.

Show that n can be properly represented and the non-divisibility condition is satisfied.

AL1.1.36
Assume the given conditions and apply the arithmetic mean – harmonic mean inequality to show that $\dfrac{5n-4}{n} \geq n$.

Factor this inequality to be equivalent to a product less than or equal to 0.

Find the only integer possibilities for n.

Consider each possibility for n and check if it works by using properties of integers and numeric inequalities.

AL1.1.37
Multiply the left-hand member by $a(b+c) + b(c+a) + c(a+b)$ and apply Cauchy inequality.

Show that $\dfrac{1}{a} + \dfrac{1}{b} + \dfrac{1}{c} = ab + bc + ca$ using the hypothesis.

Replace this in the right-hand member of the obtained Cauchy inequality.

Show that the left-hand member of the inequality to prove is larger than or equal to $\dfrac{ab+bc+ca}{2}$.

Apply the arithmetic mean – geometric mean inequality to the right-hand member of the previous inequality and show that is larger than or equal to 3/2, using $abc = 1$.

AL1.1.38
Let (x_n) be the given sequence and let $s_n = x_1 + x_2 + \cdots + x_n$. Write the inequalities between the terms s_n equivalent to the inequalities from the hypothesis.

Starting with $0 < s_{11} < s_4 < ...$ write an inequality chain holding 17 terms s_n and finishing with $s_7 < 0$, contradiction.

Show that the initial sequence cannot have 17 terms.

Show that 16 is the answer, by taking a non-contradictory inequality chain starting with $s_{10} < s_3 < ...$.

AL1.1.39
Write x^y as a product of $y + 1$ factors, from which one is 1.
Apply the arithmetic – geometric mean inequality to this product and show that $x^y \le \left(\dfrac{yx+1}{y+1}\right)^{y+1}$.

Add and subtract x in the numerator of the fraction in the right-hand member and factor out $y + 1$ in order to show that the right-hand member is equal to the right-hand member of the given equation.

Study the case in which the equality holds in the above inequality.

AL1.1.40
Show that we cannot have $x = y$ or $x = y + 1$, so $x \ge y + 2$.
Study the cases $y = 1$ and $y = 2$.
Show by induction on x that if $y \ge 3$ then $C_x^y > x + y$.

AL1.1.41
Write numbers N and $(n+1)^2$ in their decimal notation.
Subtract the relations side by side and express $2n + 1$.
Show that the difference between the second and the third digit of N is odd.
Show that $10 \le n+1 \le 31$.
Find the bounds of $2n + 1$ from the above double inequality.
Show that n can be only 13 or 22.
Check both cases.

AL1.1.42
Assume that m and n exist in the given conditions and $m \le n$.
Show that $n^2 + 4m \ge (n+1)^2$.
Show that $n^2 + 4m \le (n+2)^2$.
Show that $4m = 2n + 1$, contradiction.

AL1.1.43
Give values to n from 2 to n in the recurrence relation and add together the obtained equalities.

Put in evidence $\sum_{i=2}^{n-2} x_i$ and factor out $p + q - 1$.

Show that $\sum_{i=2}^{n-2} x_i > 0$ and find the sign of the right-hand member.

In the above equality add in both sides $(p + q - 1)(x_0 + x_1 + x_{n-1} + x_n)$.

Put in evidence $\sum_{i=0}^{n} x_i$ and write its expression as a fraction with denominator $p + q - 1$.

Show that $\sum_{i=0}^{n} x_i$ is integer and deduce that $E \vdots p + q - 1$.

AL1.1.44
Show that $x = 1$ using that the number of digits of the multiples of N is constant.

Show that no digit can be 0.

Show there is the following order relation between the digits: $x < z < y < u < v < t$.

Write the decimal notations of numbers N, $2N$, $3N$, $4N$, $5N$, $6N$.

By using the divisibility of each expression by the multiplier of N, show that: y can be 2, 4, 6, or 8; $x + y + z + t + u + v$ is multiple of 3; $10z + t$ is multiple of 4; $u = 5$; z can be 2, 4, 6, or 8.

From these conditions find the value of each digit.

AL1.1.45
If $a_i, i = \overline{1, n}$ are the numbers of sequence M, we consider the partial sums of the first 1, 2, 3, ..., n terms.

If one of these sums is divisible by n, determine the searched subsequence.

If none of the partial sums is divisible by n, find the number of their reminders upon division by n.

In the opposite case, apply the counting principle to show that there are two partial sums that give the same remainder upon division by n.

Write the difference of the two sums for determining the terms of the searched subsequence.

AL1.1.46
Write the possible cases for the positiveness of a and b resulting from $ab > 0$. Consider the case $a > 0$, $b > 0$.

Write the expression of c and replace it back in the given inequalities.

Divide the obtained inequalities by a and b (separately).
Show that $a \geq b$ and $b \geq a$, hence $a = b$.
Show that $c = a = b$ using the expression of c.
Solve the case $a < 0$, $b < 0$ analogously.

AL1.1.47
Show that f admits an inverse function $f^{-1} = f$.

Assume by absurdum that for any $x \in E$, $x \neq f(x)$. Using the bijectivity of f and that the domain of f equals to its range, show that the number of elements of E is even, contradiction.

AL1.1.48
Take $n = 1$ in the given relation and show that $a_1 = 1$.

For $n = 2$, show that $a_2 = 2$.

Prove by induction that $a_n = n$:

Make the replacement $n \to n+1$ in the given relation.

Use the formula of the sum of the first n natural numbers and the formula of the sum of cubs of the first n natural numbers to obtain a quadratic equation in a_{n+1}.

Show that the only positive solution of this equation is $a_{n+1} = n+1$.

AL1.1.49
Consider the case n is even. Write $n = 2m$ and show that the given number is of form $5c + 2 \cdot (-1)^m$.

Show that the last digit of the given number can only be $\{2, 3, 7, 8\}$.

Check that none of the perfect squares can have the last digit in the above set.

Consider the case n is odd and write $n = 2m + 1$. Show that the given number is of form $3p - 1$.

Reductio ad absurdum: assume the given number is perfect square. Show that the given number is of form $(3q\pm 1)^2$.

Show that the given number is of form $3t + 1$.

Show that the number cannot be simultaneously of forms $3p - 1$ and $3t + 1$, contradiction.

AL1.1.50
Show that it suffices to prove the affirmation for n natural number, using the formula $a^{-n} = \dfrac{1}{a^n}$.

Prove by induction:

Check if the sentence holds true for $n = 1$, $n = 2$ and $n = 3$:

Write the expression of the number, putting in evidence $a + \dfrac{1}{a}$.

Prove the sentence for $n + 1$, putting in evidence $a^n + \dfrac{1}{a^n}$ and $a + \dfrac{1}{a}$ in the expression of $a^{n+1} + \dfrac{1}{a^{n+1}}$.

AL1.1.51
Consider the case $x = y$. Reductio ad absurdum.

Show that x and z are divisible by 2 ($z = 2k$, $x = 2q$).

Replace this back in the equation and prove that k and q are also divisible by 2.

Repeat the procedure for the quotients of k and q and so on.

Show that the built sequence of quotients stops at 1, which is a contradiction.

Consider the case $y > x$. Reductio ad absurdum.

Show that $z \geq y + 1$.

Express x^n in terms of z^n and y^n and factorize.

Use $z \geq y + 1$, $z > x$ and $y > x$ to show that $x > n$, contradiction.

AL1.1.52
Consider the non-trivial case $a_0 > 1$ (the first term).

Consider the given integers a and b in the progression such that a^j and b^k are congruent with a_0 modulo p (the common difference).

Using that j and k are relatively prime, show that there exist positive integers r and s such that $jr + ks \equiv 1 \pmod{p-1}$.

Consider the number $(a^s b^r)^{jk}$ and find a power of a_0 that is congruent with it modulo p.

Apply Fermat's little theorem for this last power of a_0.

AL1.1.53
Write the formula of the area of the triangle in terms of $\sin\theta$ (the angle between the sides with lengths p and q).

Write the Pitagora's generalized theorem for the opposite side of that angle.

Assume an order between p and q.

Show that q cannot be even, starting from the assumption $q = 2$ and using the inequalities between the side-lengths of the triangle.

Using that $\sin\theta$ and $\cos\theta$ are rational, write them as fractions with two integer variables a and b, in expressions that hold powers of 1 and 2 of a and b.

Show that the only possibilities are $a^2 + b^2 = pq$, $a^2 + b^2 = p$ or $a^2 + b^2 = q$.

Study the possible numbers of solutions of each of the three equations, which give the possible value of x.

AL1.1.54
Reductio ad absurdum: assume the divisibility.

Suppose an order ($a > b$) and choose the smallest a with the given properties.

Write the theorem of division for a and b.

If q is the quotient and r the remainder, show that $1 + ab$ divides $a(b - r - 1) + b^2 + a - q + 1$.

Show that this quantity is positive and less than $2(1 + ab)$.

Show that $(1 + br) \mid (b^2 + r^2)$.

Show that $r \mid b$.

Show that the last two relations contradict the fact that a and b are relatively prime.

AL1.1.55
Reductio ad absurdum.

Show that $A \neq \phi$ and consider x in A.

Starting from $x + n \notin A$, show that $x + 2n \in A$.

Repeat the argument to show that $x + mn \in A \Leftrightarrow m$ is even.

Starting from $x + n + m \in A$, show that $x + 2m \in A$.

Repeat the argument to show that $x + nm \in A \Leftrightarrow n$ is even.

AL1.1.56
Apply Ptolomeu's inequality to diagonals d_1 and d_2.

Multiply the two given inequalities together.

Apply Cauchy-Buniakowsky-Schwartz inequality to the right-hand member.

Reductio ad absurdum. Show that $pq > d_1 d_2$, contradiction.

AL1.1.57
Denote the given sum by $S(n)$, show that $S(1)$, and then $S(n + 1) = S(n)$:

Write $k = n + 1 + h$ and rewrite the given expression.

Show that $2^{n+1} S(n+1) = 2^n S(n) + 2^n S(n+1)$, using $C_{n+1+h}^{n+1} = C_{n+1+h}^{h} = C_{n+h}^{h} + C_{n+h}^{h-1}$ and splitting some sums.

AL1.1.58
Let A be a subset of the given set. For each $a \in A$, denote by a_0 the odd part of a (so that $a = a_0 2^i$ with a_0 odd and i a non-negative integer).

Write A as a partition of A_m sets, where A_m is the set of $a \in A$ with $a_0 = m$, for each odd integer m.

Consider three ranges of m, with heads in 32, 64, 128 and 256.

For each range, find an upper bound of the cardinality of A_m using the double-free condition.

Add up these bounds to find an upper bound for the cardinality of A. This is 171.

Show that this bound can be achieved, studying when the inequalities found in each range of m for $|A_m|$ become equalities.

AL1.1.59

Assume P is a polynomial satisfying (a), (b), and (c).

Consider the polynomial $Q(x) = P(x) - x^{11}$, study its degree and write it in its factorial form using (a).

Determine the constant from the factorial form using (c).

Show that $n + 1$ is a divisor of 2004 using (b).

Find the smallest number $n > 11$ such that $n + 1$ divides 2004, factoring 2004.

AL1.1.60

Let $(a_i)_{i=1}^{2006}$ the sequence of 2006 terms from D and let $P_n = \prod_{i=1}^{n} a_i$, $P_0 = 1$. Show that any block of consecutive terms from $(a_i)_{i=1}^{2006}$ is of the form P_n / P_m, with $0 \leq m < n \leq 2006$.

Show that, for a suitable choice of (m, n), P_n / P_m is a perfect square:

Show that $P_n = \prod_{i=1}^{10} d_i^{\alpha_{in}}$ with α_{in} non-negative integers.

Evaluate P_n / P_m.

Show that P_n / P_m is a perfect square if the numbers $\alpha_{kn} - \alpha_{km}$, $k = 1, \ldots, 10$, are all even.

Take the vector $\alpha_n = (\alpha_{1n}, \ldots, \alpha_{10n})$ and show that there are 1024 possible parity combinations for its components.

Using that there are 2007 such vectors, apply the counting principle to show that two of these must have the same parity combination.

AL1.1.61

Assume x is a solution and show that $[x]$ is of form p^2 with p natural number.

Conversely, show that if $[x]$ is of form p^2, then the given equation is satisfied:

Show that x is of form $p^2 + \varepsilon$ with $\varepsilon \in [0, 1)$.

Show that $\sqrt{p^2 + \varepsilon}$ lies between p and $p + 1$, using the inequalities $\varepsilon < 1$ and $p^2 + 1 \leq (p+1)^2$.

Write the solution of the given equation as an infinite union of intervals.

Generalization: n-th root, $n > 2$.

By a similar reasoning, show that x is of form $p^n + \varepsilon$, with $\varepsilon \in [0, 1)$.

AL1.1.62

Show that n is blocky if and only if $n > 1$ and $n \equiv \pm 1$ or $n \equiv \pm 7$ (mod 24):

Find integers k, n, and m such that $m^2 = \dfrac{1}{n} \sum_{i=k}^{k+n-1} i^2$:

Write the above sum as a difference of sums indexed from $i = 1$.

Using the standard formula $\sum_{i=1}^{N} i^2 = N(N+1)(2N+1)/6$, show that $m^2 = k^2 + (n-1)k + (n-1)(2n-1)/6$.

Show that $n \equiv \pm 1$ (mod 6), using the relation between the integers above.

Complete a square on the right side of the above relation.

Evaluate $(n^2 - 1)/12$ as a product of two linear expressions in m, k, and n, writing it first as a difference of two squares.

Show that $(n-1)/2$ is of form $3l$ or $3l - 1$ and evaluate $(n^2 - 1)/12$.

Show that $(n^2 - 1)/12 \equiv 0$ (mod 4), using the parity.

Show that $n \equiv \pm 1$ or $n \equiv \pm 7$ (mod 24).

Conversely, assuming n is of this form, show that $(n^2 - 1)/12 = (2\alpha)(2\beta)$.

Take $m = \alpha + \beta$ and $k = \alpha - \beta - (n-1)/2$ and verify the initial relation.

Take all possible factorization of $(n^2 - 1)/12 = (2\alpha)(2\beta)$ and set $k = \alpha - \beta - (n-1)/2$. This gives all possible starting points k for the summation.

Study the construction above and show that the number of such k is the number of positive divisors of $(n^2 - 1)/48$.

AL1.1.63

Assume $C(a, b)$ is a center of symmetry of the graph of P. Write the relation between $P(a - h)$ and $P(a + h)$.

Substitute $y = x - a$ in the above relation and let $R(x)$ be a polynomial such that $P(x) = R(y)$.

Take $x = a - h$ and $x = a + h$ and replace them in the first relation, putting in evidence polynomial R.

Write $R(h)$ in its standard unfolded form, with coefficients a_i and degree n.

Replace in the first relation $R(h)$ and $R(-h)$.

Identify the coefficients and show that $a_0 = b$ and $a_2 = a_4 = a_6 = \cdots = 0$.

Write $R(h)$ as $b + hQ(h^2)$.

Write $P(x)$ as a function of R.

For the sufficiency, assume the existence of Q in the given conditions. For any real h, make the substitutions $x = a - h$ and $x = a + h$.

Show by calculation that $P(a - h) + P(a + h) = 2b$.

AL1.1.64

Put in evidence $x^2 + 1$ in the numerator.

Write the expression of the given function by putting in evidence the variable $a = \sqrt{x^2 + 1}$.

Show that $a > 0 \Rightarrow a + \dfrac{1}{a} \geq 2$ and prove that $f(R) \subset [2, +\infty)$.

Prove the inverse inclusion, considering $u \in [2, +\infty)$ and solving the equation $f(x) = u$ (by raising to the second power you will obtain a biquadratic equation in x).

AL1.1.65

Study the cardinalities of the two sets and show that if a bijection exists then $n^2 = 2^n$, where n is the cardinality of A.

Solve in natural numbers the equation $n^2 = 2^n$:

Check it for $n = 1, 2, 3, 4$.

Show that $n^2 > 2n + 1$ for $n \geq 3$

Argue by induction that if $n \geq 5$ then $2^n > n^2$, using the above inequality.

Calculate the number of bijections (as the factorial of number of elements) for cases $n = 2$ and $n = 4$.

AL1.1.66

Let $y_i, i = \overline{1, n}$ n strictly positive numbers. Apply the arithmetic mean – geometric mean inequality for numbers $\dfrac{y_1}{y_2}, \dfrac{y_2}{y_3}, \ldots, \dfrac{y_n}{y_1}$ and show that $\dfrac{y_1}{y_2} + \dfrac{y_2}{y_3} + \cdots + \dfrac{y_n}{y_1} \geq n$.

Replace one at a time numbers y_i with $x_i^n, x_i^{n-1}, \ldots, x_i^2, x_i$ and write the resulting inequalities.

Multiply the first inequality by a_0, the second by a_1, and so on, the n-th by a_{n-1}, where a_i are the coefficients of polynomial P.

Add together the obtained inequalities.

Add na_n in both sides.

Show that $P\left(\dfrac{x_1}{x_2}\right) + P\left(\dfrac{x_2}{x_3}\right) + \cdots + P\left(\dfrac{x_n}{x_1}\right) \geq nP(1)$.

In the Cauchy-Buniakovsky-Schwarz inequality, set all the terms in the first sum equal to 1 and the terms in the second sum equal to $P\left(\dfrac{x_1}{x_2}\right), P\left(\dfrac{x_2}{x_3}\right), \ldots, P\left(\dfrac{x_n}{x_1}\right)$.

Use the inequalities deduced above to show the required inequality.

AL1.1.67
Study the case $E(a) = 0$.

Show that $E(-a) = -E(a)$. Determine the values of expression $E(a)$ for $a > 0$:

Write $E(a)$ as a fraction, using the formula of difference between two squares and show that $0 < E(a) < 1$.

Take $b \in (0, 1)$ and show that equation $E(a) = b$ has a solution $a > 0$.

Write the interval of values of expression $E(a)$ by making the union of the intervals found for $a > 0$ and $a < 0$.

AL1.1.68
Substitute the three radicals with three new variables.
Group the terms conveniently and complete three squares.
Write the equation as a sum of three squares that is equal to 0.
Find the solution, equaling each square to 0.

AL1.1.69
Show that $7n^2 - m^2 > 0$.

Find the least $k \in N^*$ such that $7n^2 - m^2 = k$, considering all possible forms of number m regarding r, where $m = 7q \pm r$.

For this k, put the condition $7n^2 - m^2 \geq k$ and prove the required inequality.

AL1.1.70
Prove by induction that $f(a_k) = a_k$ for $k = 1, \ldots, n$:

For $n = 2$, write the two possible cases of correspondence of function f, using its bijectivity. Show that the case where $f(a_1) = a_2$ is impossible.

Assume the affirmation true for $n - 1$ numbers. Prove the affirmation for n numbers:

Study the case $f(a_n) = a_n$.

Assume by absurdum that $f(a_n) \neq a_n$. Show that there exists $p < n$ such that $f(a_p) = a_n$.

Use the given inequalities for showing that $a_p < f(a_{p+k})$,

$k = 0, 1, 2, \ldots, n-p$ and there exist at least $n-p+1$ numbers larger than a_p, which contradicts with hypothesis ($n-p$ numbers).

Find a counterexample for the situation where A is the set of integer, taking a function that preserves the order.

AL1.1.71
Choose a order between the three primes.

Express q and observe that it is the arithmetic mean of three numbers, from which two are among numbers p_i.

Show that q is between two of the three primes.

Show that q cannot be equal to the third prime, using that the given primes are odd.

AL1.1.72
Raise to the second power in the given relation.

Use Pitagora's theorem for express a and a'.

Show that $bb' = cc'$: Clear the identical terms and group the remaining terms to obtain a square equal to 0.

Show that $\dfrac{b}{b'} = \dfrac{c}{c'}$.

Show that $b = c$ and $b' = c'$.

AL1.1.73
Show that number x has exactly two digits: assume it has exactly one digit and arrive at a contradiction, studying the maximum of expression $x + y + z$; assume that x has exactly three digits and arrive at a contradiction, studying the minimum of expression $x + y + z$.

Write $x = 10a + b$ and study the cases $a + b \leq 9$ and $9 < a + b \leq 18$. In each case, find a relation between a and b.

Build tables of values for a and b in both cases and choose from them the possible values (a, b).

AL1.1.74
Express $\left[\sqrt{n}\right]$ and show that n is of form $3k + 1$, $k \in N^*$.

Replace n in the expression of $\left[\sqrt{n}\right]$.

From the double inequality of the definition of integer part, deduce that $k > 1$ and $\dfrac{3-\sqrt{13}}{2} \leq k \leq \dfrac{3+\sqrt{13}}{2}$.

Show that the only possible values are $k = 2$ and $k = 3$.

AL1.1.75
Assume f is not injective and consider two distinct natural numbers u and v such that $f(u) = f(v)$.

Express $f(u) - f(v)$ using the definition of fractional part and factor out $m^u - m^v$.

Express $\sqrt{2}$ from the above relation and deduce that it is rational, contradiction.

AL1.1.76
Show that $\prod_{i=1}^{n} x_i \in (0, 1)$, using the properties of the logarithm and the first given inequality.

Show that at least one number x_k is smaller than 1.

Show that there exist an even number of numbers x_i smaller than 1, using the properties of the logarithm and the second given inequality.

Express $2s_n$, writing the logarithms of products as sums of logarithms and using the formula of the square of a sum $\left(\sum_{i=1}^{n} y_i\right)^2$.

AL1.1.77
Show that all numbers $\sqrt{m}$, with m prime, are irrational.

Consider all numbers $\sqrt{m}$, with m prime, and denote them by $y_1, y_2, \ldots, y_m, \ldots$. Build the infinite table $(y_i x_j)_{i=1,\infty}^{j=1,n}$ with n columns and an infinity of rows. Show that there exists a row containing only irrational numbers:

Prove by reductio ad absurdum: Assume that each row holds at least one rational number. Show that there exists at least one column holding two rational numbers.

Show that the ratio of these two numbers is irrational, contradiction.

AL1.1.78
Show that all terms from the fourth upward are multiples of 10. Show that the last digit of N is 7.

Reductio ad absurdum: Assume N is a perfect square. Show that the last digit of N cannot be 7, contradiction.

AL1.1.79
Show that $5^{7^n} = 5^{3k+1}$, writing $7 = 6 + 1$.
Show that $5^{3k+1} = (31m+1)^k \cdot 5$, writing $125 = 124 + 1$.
Show that $5^{7^n} = 31q + 5$.

AL1.1.80
Denote by $f(x, y)$ the expression within the modulus. Express $f(x, y)$, writing $\sin 2x - \sin 2y$ as a product and factoring out $t = \sqrt{1 + \sin^2(x - y)}$.

Show that there exists $g(x, y)$ real such that $\sin g(x, y) = \dfrac{\sin(x - y)}{t}$ and $\cos g(x, y) = \dfrac{1}{t}$.

Replace these expressions in the expression of $f(x, y)$ and show that $|f(x, y)| \leq 2\sqrt{2}$, using $|\sin a| \leq 1$, $\forall a \in R$.

AL1.1.81
Show that there exists at least one odd number by reductio ad absurdum: Assume all numbers are even and study the minimum of their sum. Show that this minimum is higher than the given sum, contradiction.

Assume there is only one odd number among them and arrive at a contradiction regarding the parity of the sum.

AL1.1.82
Study the case $x \geq 2$, $y \geq 2$ and show there are no solutions due to the different parities of the two members.

Study the case $x = 0$ and show there are no solutions.
Study the case $x = 1$ and show that the only solution is $y = 2$.

Use the symmetry of the given equation to find all solutions.

AL1.1.83
Show that $x > 0$, using the properties of the exponential function.

Apply the arithmetic mean – geometric mean inequality for $x \cdot 2^{\frac{1}{x}}$ and $\frac{1}{x} \cdot 2^x$.

Show that $\frac{1}{x} + x \geq 2$.

Show that $x \cdot 2^{\frac{1}{x}} + \frac{1}{x} \cdot 2^x \geq 4$ and study the case when the equality holds.

AL1.1.84
Show that N is divisible by 9 and 25:
Show that N is divisible by 25, using that $\underbrace{11\ldots10}_{n}$ is divisible by 5.

Write N as a product, using the formula of difference between two squares.

In the first factor of the product, write $n = \underbrace{1+1+\cdots+1}_{n}$.

Write the first factor as a sum of terms of form $10^k - 1$.
Show that the first factor is divisible by 9.

AL1.1.85
Express $B - 2A$ and factor out $(p - n)$.
Show that $B - 2A$ is divisible by $(p - n)(a - b)$.
Show that B is divisible by $(p - n)(a - b)$.

AL1.1.86
Consider the height corresponding to the side of length b. Write the formula of area of the given triangle.

Use the fact that the height is smaller than the adjacent sides and the given inequalities.

Add together the two obtained inequalities.

Assume the triangle is right-angled. Use the two inequalities obtained when proving the first required inequality and the fact that $n^2 < 2n^2$.

Add together the two obtained inequalities and use Pitagora's theorem.

AL1.1.87
Express a_{n+2} in terms of a_n and a_{n-1}.

Express a_{n+4} and show that is equal to $-a_n$.

Show that the sequence is periodic of unique period 8, expressing a_{n+8}.

Show that the sequence is bounded.

AL1.1.88
Show that N_2 is a perfect square, writing it as a square of a trinomial.

Prove that the number of divisors of the square of a natural number is odd:

Argue by contradiction that if $N = d_1 d_2$ and $d_1 < \sqrt{N}$, then $d_2 > \sqrt{N}$.

Show that $x^2 + 1$ cannot be a perfect square, bounding it with two consecutive squares.

Show that p is even and q is odd.

AL1.1.89
Assume k natural and study the problem for k integer.

Represent numbers $k = 0, 1, 2, 3$ in the required form.

Show that $(m+1)^2 - (m+2)^2 - (m+3)^2 + (m+4)^2 = 4, \forall m \in N$.

Prove the existence by complete induction:

Assume $k = \pm 1^2 \pm 2^2 \pm \cdots \pm m^2$. Show that $k + 4$ can be represented in this form, using the above identities.

In identity $(m+1)^2 - (m+2)^2 - (m+3)^2 + (m+4)^2 = 4$, replace m with $m + 4$.

Show that
$$(m+1)^2 - (m+2)^2 - (m+3)^2 + (m+4)^2 - (m+5)^2 + (m+6)^2 +$$

$+(m+7)^2 - (m+8)^2 = 0, \forall m \in N$.

Show that from the representation of k in form $k = \pm 1^2 \pm 2^2 \pm \cdots \pm m^2$ one can obtain an infinity of such representations, replacing m with $m + 8$, $m + 16$, etc.

AL1.1.90

Show that the first required inequality is equivalent to $\left(1+n^{-1}\right)^{nb} \geq b^d$, grouping the logarithms in one side and multiplying both sides by n.

Show that $\left(1+\dfrac{1}{n}\right)^n \geq 2, \forall n \in N^*$.

Write the first required inequality for $k = 1, \ldots, n$ and add together the n obtained inequalities.

Transform the sums of logarithms into logarithms of factorials.

Prove the second required inequality by induction: Check it for $n = 2$.

Assume the inequality true for n. Show it is true for $n + 1$, using the inequality obtained after adding together the n inequalities and the induction hypothesis:

Show that $\log_b((n+1)!) > \dfrac{d(n+1)}{b} \sum_{k=2}^{n+1} \dfrac{1}{k}$:

Split the sum $\sum_{k=1}^{n} \dfrac{1}{k}$ into $\sum_{k=2}^{n} \dfrac{1}{k}$ and 1.

Factor out $\sum_{k=2}^{n} \dfrac{1}{k}$ and $\dfrac{d}{b}$ separately after you group conveniently the terms.

Write the free term $\dfrac{d}{b}$ as $\dfrac{d(n+1)}{b(n+1)}$.

AL1.1.91

Substitute each factor from the left-hand member with a power of 3 (with exponents u, v, t, such that $u + v + t = y$).

Write the relations between $3^u, 3^v, 3^t$.

Factor out in these relations.

Show that one of the exponents u, v, t is 0 and find the other two unknown values.

Find x and y.

AL1.1.92

Express a_{n+2} in terms of a_n by using the given recurrence relation.

Show that $a_n \neq 1$.

Express a_{n+4} in terms of a_{n+2}, using the new recurrence relation obtained previously.

Show that $a_{n+4} = a_n$, $\forall n \in N$.

AL1.1.93

Study the case $x > 0$:

Show that $x^{2n} < \sum_{i=0}^{2n} x^i < (x+1)^{2n}$.

Use the given relation and show that $x < y < x + 1$, impossible in integers.

Study the case $x < -1$:

Show that $(x+1)^{2n} < \sum_{i=0}^{2n} x^i < x^{2n}$:

Write $\sum_{i=0}^{2n} x^i = \dfrac{1-x^{2n+1}}{1-x}$. Denote $1 + x = -a$, $a > 0$.

Show that $a^{2n} < \dfrac{1+(1+a)^{2n+1}}{2+a}$, using the binomial theorem.

Deduce that $x + 1 < y < x$, impossible in integers.

Study the cases $x = 0$ and $x = -1$ and find the solutions of the given equation.

AL1.1.94

Write u_n and v_n in terms of x_n and y_n.

Replace back $u_{n+1}, v_{n+1}, u_n, v_n$ in the given relations and show that $x_{n+1} = 3x_n + 2y_n$, $y_{n+1} = 4x_n + 3y_n$.

Calculate x_0, y_0.

Show that all y_n are positive integers.

Express $y_{n+1}^2 - 2x_{n+1}^2$ and show that it is equal to $y_n^2 - 2x_n^2$.
Show that $y_n^2 - 2x_n^2 = -1$ for all n.
Express $\left[\sqrt{2}x_n\right]$ from the above relation.
Show that $\left[\sqrt{y_n^2+1}\right] = y_n$.

AL1.1.95
Check that polynomials $P = 0$ and $P = 1$ satisfy the given condition.

Assume $P \neq 0$, $P \neq 1$. Show that P is of the form $a_n X^n$ by reduction ad absurdum:

Write P in its standard form, with coefficients a_i and degree n.

Assume that there exists $k \in \{0, 1, \ldots, n-1\}$ such that $a_k \neq 0$ and $a_s = 0$ for any $s \in \{k+1, k+2, \ldots, n-1\}$. Write the given relation by replacing the coefficients according to this assumption.

Study the terms in x^{n+k} in both sides and show that such a term does not exist in the left-hand member, contradiction.

Write the given equality for $P = a_n X^n$ and show that $a_n = 1$.

AL1.1.96
Show that the possible remainders of a perfect square upon division by 7 are 0, 1, 2, 4, using the theorem of division.

Consider five integers and their squares.

Apply the previously proven property to these squares.

Show that at least two of these squares have the same remainder upon division by 7, using the counting principle.

Express the difference of these two squares and show that the sum or the difference of their square roots is divisible by 7.

AL1.1.97
For fixed n, consider sets $A = \{n, n+1, \ldots, 2n\}$, $B = \{n! + n, n! + n + 1, \ldots, n! + 2n\}$ and function $f : A \to B$, $f(x) = x + n!$.

Show that f is bijective.
Show that if $x \in A$ is composite, then $f(x) \in B$ is composite.

Show that B has at most as many primes as A has.

Write the expressions of the numbers of primes from the two sets in terms of n and π and express the previous inequality.

AL1.1.98

Express $P_{n,m}\left(\dfrac{1}{x}\right)$ in terms of $P_{n,m}(x)$.

Identify the coefficients in the above relation and show that $a_{n+m-j} = -a_j,\ \forall j \in \{0, 1, \ldots, n+m\}$.

Show that $a_{\frac{n+m}{2}} = 0$.

Show that for every index $j \leq \dfrac{n+m}{2} - 1$ for which $a_j = 0$ there exists an index $k \geq \dfrac{n+m}{2} + 1$ for which $a_k = 0$.

Show that $c_{n,m}$ is odd.

Show that for $n = m$, $P_{n,n}$ contains only even powers of x.

Show that $c_{n,n} = n \geq 3$.

Study the counterexample $n = 7,\ m = 15$:

Write $P_{7,15}(x)$ by using the formula of the difference of two squares and developing the powers of the two binomials.

Show that $a_3 = 0$.

Show that $a_{19} = 0$ and $a_{11} = 0$ by using the proof of point a).

Show that $c_{7,15} \geq 3$.

AL1.1.99

Write the relation of the theorem of division for the division of k^p by p^2.

Show by reductio ad absurdum that $a_k = 0$ is impossible.

Write the relation of the theorem of division for the division of $(p-k)^p$ by p^2.

Develop $(p-k)^p$ by the binomial theorem.

Using the previous development, show that $k^p + (p-k)^p$ is multiple of p^2.

Show that $a_k + a_{k-p}$ is multiple of p^2.

Show that $a_k + a_{k-p} = p^2$.

Write the previous relation for every k from 1 to $p-1$ and add together these equalities.

AL1.1.100

Add and subtract $2n^2\sqrt{a}$ in the expression of P.
Group the terms to complete a square.
Write P as a difference of two squares and factorize it.
Choose $\sqrt{2\sqrt{a}} = k \in N$.
Choose $k = 2m$, $m \in N$.
Replace back these values in the expression of P.
Show that if $m > 1$, then each of the two factors in the expression of P is bigger than 1.

Show that the terms of the sequence $\left(2m^2\right)^2$ represent values of a with the given property.

AL1.1.101

Check that the implication "$\Leftarrow$" is obvious ($P(x) = P(-x)$).

For the implication "$\Rightarrow$" assume function P is even and write the relation $P(x) = P(-x)$ using the standard form of polynomial P, with coefficients a_i and degree n.

Show that all coefficients of the terms with odd powers of x are null.

Show that P can be written as $P(x) = P_1(x^2)$, for any real x.

Consider numbers $y_i \in C$ such that $y_i^2 = x_i$ ($i = 1, ..., n$) and $b \in C$ such that $b^2 = (-1)^n a$ and rewrite the relation obtained previously.

Group the factors conveniently, use the formula of difference between two squares and express $P(x)$ in the form $Q(x)Q(-x)$.

PROOFS

AL1.1.1

We add together the two given equations. After simplifying the resulting equation and completing the squares, we arrive at the following equation: $(x-5/2)^2 + (y-5/2)^2 = 1/2$. (1)

We subtract the second given equation from the first and group the terms:
$$xy(y-x) + 6(x-y) + (x+y)(x-y) = xy(x-y) + (y-x)$$
$$(x-y)[-xy + 6 + (x+y) - xy + 1] = 0$$
$$(x-y)(x+y-2xy+7) = 0$$

Thus, either $x - y = 0$ or $x + y - 2xy + 7 = 0$. The only ways to have $x - y = 0$ are with $x = y = 2$ or $x = y = 3$ (found by solving equation (1) with the substitution $x = y$). Now, all solutions to the original system where $x \neq y$ will be solutions to $x + y - 2xy + 7 = 0$. This equation is equivalent to the following equation (derived by rearranging terms and factoring): $(x-1/2)(y-1/2) = 15/4$. (2)

Now we solve equations (1) and (2) simultaneously.

Let $a = x - 5/2$ and $b = y - 5/2$. Then, equation (1) is equivalent to $a^2 + b^2 = 1/2$. (3)
and equation (2) is equivalent to:
$$(a+2)(b+2) = 15/4 \Rightarrow ab + 2(a+b) = -1/4$$
$$\Rightarrow 2ab + 4(a+b) = -1/2. \quad (4)$$

Adding equation (4) to equation (3), we find:
$$(a+b)^2 + 4(a+b) = 0 \Rightarrow a+b = 0, -4. \quad (5)$$

Subtracting equation (4) from equation (3), we find:
$(a-b)^2 - 4(a+b) = 1$. (6)

Observe that if $a + b = -4$, then equation (6) will be false. Thus, $a + b = 0$. Substituting this into equation (6), we obtain:
$$(a-b)^2 = 1 \Rightarrow a - b = \pm 1 \quad (7)$$

Since $a + b = 0$, we now can find all ordered pairs (a, b) with the help of equation (7). They are $(-1/2, 1/2)$ and $(1/2, -1/2)$.

Therefore, the only solutions (x, y) are $(2, 2)$, $(3, 3)$, $(2, 3)$, and $(3, 2)$.

AL1.1.2
Assume that for some $k > 3$, $a_k = 1997$. Then, each of the four numbers $a_{k-1}, a_{k-2}, a_{k-3},$ and a_{k-4} must exist.

Let $w = a_{k-1}$, $x = a_{k-2}$, $y = a_{k-3}$, and $z = a_{k-4}$.

By the given condition, $1997 = w^2 + x^2 + y^2$. Thus, $w \le \sqrt{1997} < 45$ and since w is a positive integer, $w \le 44$.

But then $x^2 + y^2 \ge 1997 - 44^2 = 61$.

Now, $w = x^2 + y^2 + z^2$. Since $x^2 + y^2 \ge 61$ and $z^2 \ge 0$, $w = x^2 + y^2 + z^2 \ge 61$. But $w \le 44$, contradiction.

AL1.1.3
The smallest such a is 6. The equation holds if $a = 6$, $x = 8$.
We write the equation as

$$[\cos \pi(a-x) - 1]^2 + \left[\cos\frac{3\pi x}{2a} \cos\left(\frac{\pi x}{2a} + \frac{\pi}{3}\right) + 1\right] = 0.$$

Since both terms on the left side are non-negative, equality can only hold if both are 0.

From the first term we get that x is an integer congruent to a (mod 2). From the second term we see that each cosine involved must be -1 or 1 for the whole term to be 0.

If $\cos\left(\frac{\pi x}{2a} + \frac{\pi}{3}\right) = 1$ then $\frac{\pi x}{2a} + \frac{\pi}{3} = 2k\pi$ for some integer k, and multiplying through by $\frac{6a}{\pi}$ gives $3x \equiv -2a \pmod{12a}$, while if the cosine is -1 then $\frac{\pi x}{2a} + \frac{\pi}{3} = (2k+1)\pi$ and multiplying through by $\frac{6a}{\pi}$ gives $3x \equiv 4a \pmod{12a}$. In both cases we have $3x$ divisible by 2 and hence so is a.

Also our two cases give $-2a$ and $4a$, respectively, are divisible by 3, so a is divisible by 3.

Therefore $6 \mid a$ and our solution is minimal.

AL1.1.4

We have that $a^2 - b$, $b^2 - c$, and $c^2 - a$ are perfect squares. Since $a^2 - b$ is a perfect square smaller than a^2, we have $a^2 - b \leq (a-1)^2$, which is equivalent to $b \geq 2a - 1$.

Likewise, $c \geq 2b - 1$ and $a \geq 2c - 1$.

Adding these three inequalities together gives $a + b + c \leq 3$, hence $(a, b, c) = (1, 1, 1)$ is the only solution.

AL1.1.5

Let $u = 2x + 3$, $v = 2y + 3$, $w = 2z + 3$. Then the given equation is equivalent to $u^2 + v^2 + w^2 = 7$.

Asking that the above equation has solutions in rational numbers is equivalent to ask that the equation $a^2 + b^2 + c^2 = 7d^2$ has non-zero solutions in integers. Assume on the contrary that (a, b, c, d) is a non-zero solution with $|a| + |b| + |c| + |d|$ minimal.

We show first that every perfect square is congruent to 0 or 1 modulo 4. Indeed, if $n = 4m + k$ with $k \in \{0, 1, 2, 3\}$, then $n^2 = 16m^2 + 8mk + k^2$ and one can easily check that for $k = 0, 1, 2, 3$, the remainders of k^2 upon division by 4 can be only 0 or 1. Hence the possible remainders of n^2 upon division by 4 are still 0, 1. Thus we proved that every perfect square has this property.

Under the condition of the solution being minimal, we have that each of a, b, c, d is congruent to 0 modulo 4.

Thus, we must have a, b, c, d even, therefore d is also even.

But then $(a/2, b/2, c/2, d/2)$ is also a solution of our equation and is a smaller solution, which is a contradiction.

AL1.1.6

The last digits of n and n^5 are the same. Hence, the last digit of the left-hand side is the same as that of $3 + 0 + 4 + 7$, which is 4.

Hence, the last digit of k is 4. Also $133 \equiv 1 \pmod{3}$, $110 \equiv -1 \pmod{3}$, $84 \equiv 0 \pmod{3}$, $27 \equiv 0 \pmod{3}$, so the left-hand side is congruent to 0 modulo 3. Obviously, $k > 133$.

Therefore, the smallest possibility is 144, the next is 174.

Now
$11^5 = (10+1)^5 = 10^5 + 5 \cdot 10^4 + 10 \cdot 10^3 + 10 \cdot 10^2 + 5 \cdot 10 + 1 = 161051$,

so $110^5 = 11^5 \cdot 10^5 = 1.61051 \cdot 10^{10} < 2 \cdot 10^{10}$.

Obviously, 27 and 84 are smaller than 100, so 27^5 and 84^5 are smaller than 10^{10}.

Similarly, $133^5 < (1331/10)^5 = 11^{15}/10^5 < 5 \cdot 10^{10}$.

Hence, the left-hand side is smaller than 10^{11}. But $170^2 = 28900 > 28000$, $170^4 = 780000000 > 7 \cdot 10^8$ and $170^5 > 10^{11}$. Hence the only possibility for k is 144.

AL1.1.7

Let $p = \dfrac{1}{2} \cdot \dfrac{3}{4} \cdots \dfrac{1997}{1998}$ and $q = \dfrac{2}{3} \cdot \dfrac{4}{5} \cdots \dfrac{1998}{1999}$.

Note $p < q$, so $p^2 < pq = \dfrac{1}{2} \cdot \dfrac{2}{3} \cdots \dfrac{1998}{1999} = \dfrac{1}{1999}$.

Therefore, $p < \dfrac{1}{\sqrt{1999}} < \dfrac{1}{44}$. Also,

$$p = \dfrac{1998!}{\left(999! \cdot 2^{999}\right)^2} = 2^{-1998} C_{1998}^{999}, \text{ while}$$

$2^{1998} = C_{1998}^0 + \cdots + C_{1998}^{1998} < 1999 C_{1998}^{999}$. Thus, $p > 1/1999$.

AL1.1.8

From the inequality $2ab \leq a^2 + b^2$ we get

$$V_n \leq \dfrac{\sin^2 x_1 + \cos^2 x_2}{2} + \cdots + \dfrac{\sin^2 x_n + \cos^2 x_1}{2} = \dfrac{n}{2},$$

in which equality holds for $x_1 = \cdots = x_n = \pi/4$.

AL1.1.9

The only such prime is $p = 7$. Assume without loss of generality that $x, y \geq 0$. Note that $p + 1 = 2x^2$ is even, so p is odd.

Also, $2x^2 \equiv 1 \equiv 2y^2 \pmod{p}$, which implies $x \equiv \pm y \pmod{p}$ since p is odd. Since $x < y < p$, we have $x + y = p$. Then

$$p^2 + 1 = 2(p - x)^2 = 2p^2 - 4px + p + 1,$$

so $p = 4x - 1$, x is 0 or 2 and p is -1 or 7. Of course, -1 is not prime, but for $p = 7$, $(x, y) = (2, 5)$ is a solution.

AL1.1.10

The solutions are $x = y = z = t$, $t \in \left\{1, \dfrac{-1+\sqrt{5}}{2}, \dfrac{-1-\sqrt{5}}{2}\right\}$.

These are all the solutions with $x = y = z$.
Assume on the contrary that $x \neq y$.

If $x > y$, then $y = (x^3 + 1)/2 > (y^3 + 1)/2 = z$, so $y > z$, and likewise $z > x$. Hence, $y > x$, contradiction.

AL1.1.11

The only such primes are $p = 5, 17$. Adding together the two functions, we get $f(x) + g(x) = 2x^3(x+1)(x+4)$.

Thus, if p divides $f(x)$ and $g(x)$, it divides either 2, x, $x+1$, or $x + 4$.

Since $f(0) = 1$ and $f(1) = 17$, we cannot have $p = 2$.
If p divides x then $f(x) \equiv 1 \pmod{p}$, also impossible.
If p divides $x + 1$ then $f(x) \equiv 5 \pmod{p}$, so p divides 5, and $x = 4$ works.
If p divides $x + 4$ then $f(x) \equiv 17 \pmod{p}$, so p divides 17, and $x = 13$ works.

AL1.1.12

Let $k = f(x) + 1/x$. Then $k > 0$, according to (2).
According to (3), we have $f(k)f(f(x) + 1/x) = 1$.
But also $f(x)f(k) = 1$, hence
$f(x) = f(f(k) + 1/k) = f(1/f(x) + 1/(f(x) + 1/x))$.
Since f is strictly increasing, f is injective, so
$x = 1/f(x) + 1/(f(x) + 1/x)$. Solving for $f(x)$ we get $f(x) = \dfrac{1 \pm \sqrt{5}}{2x}$, and it is easy to check that only $\dfrac{1 - \sqrt{5}}{2x}$ satisfies all three conditions.

Hence $f(1) = \dfrac{1 - \sqrt{5}}{2}$.

AL1.1.13
Let $d = \gcd(x, y)$ so that $x = dx_1$, $y = dy_1$. Then the equation is equivalent to $1997 \cdot 13 y_1^2 + 1997 \cdot 1996 x_1^2 = d^2 z x_1^2 y_1^2$.

Since x_1 and y_1 are coprime we must have $x_1^2 \mid 1997 \cdot 13$ and $y_1^2 \mid 1997 \cdot 1996$.

It is easy to check that 1997 is square free, and clearly is coprime to 13 and 1996. Moreover, $1996 = 2^2 \cdot 499$, and it is easy to check that 499 is square free. Therefore, (x_1, y_1) is $(1, 1)$ or $(1, 2)$. Consider them as separate cases.

Case 1: $(x_1, y_1) = (1, 1)$. Then
$d^2 z = (13 + 1996) \cdot 1997 = 1997 \cdot 7^2 \cdot 41$. Since 1997 is coprime to 7 and 41, $d = 1$ or 7. These give respectively the solutions
$(x, y, z) = (1, 1, 4011973), (7, 7, 81877)$.

Case 2: $(x_1, y_1) = (1, 2)$. Then $d^2 z = (13 + 499) \cdot 1997 = 1997 \cdot 2^9$.
So $d = 1, 2, 4, 8$ or 16. These give respectively the solutions
$(x, y, z) = (1, 2, 1022464), (2, 4, 255616), (4, 8, 63904)$,
$(8, 16, 15976), (16, 32, 3994)$.

We also have the solutions obtained by changing the sign of x and y.

AL1.1.14
Let n be the middle integer. Then $n - 2, n - 1, n, n + 1$, and $n + 2$ are the five given consecutive positive integers.

The sum of the middle three is $3n$, which is a square m^2.
Then $3 \mid m$, so $3n = 9k^2$, hence $n = 3k^2$ for some k.

The sum of all five numbers is $5n$, which is a cube. Therefore, $15k^2$ is a cube. Hence, k is multiple of 15. The smallest value of k is 15, which gives $n = 3 \cdot 15^2$.

AL1.1.15
We claim that $f(n) = n(n + 1)!$ for $n = 1, 2, \ldots$
For $n = 1$, the equality holds trivially.
Let now $n \geq 1$ and assume that the equality holds for n.
Then
$$f(n+1) = f(n) + \left[(n+1)^2 + 1\right](n+1)! = n(n+1)! + (n^2 + 2n + 2)(n+1)!$$

$$= (n^2 + 3n + 2)(n+1)! = (n+1)(n+2)(n+1)! = (n+1)(n+2)!,$$

which proves the formula with $n + 1$ in place of n, completing the induction.

AL1.1.16

For $n = 1$, the assertion is trivially true. Suppose n is greater than 1. Since n is odd, $n - 1$ is even.

Then

$$1^n + 2^n + \cdots + n^n = \sum_{k=1}^{(n-1)/2} \left[k^n + (n-k)^n\right] + n^n$$

$$= \sum_{k=1}^{(n-1)/2} \left[k^n + n^n + C_n^1 n^{n-1}(-k)^1 + \cdots C_n^{n-1} n^1 (-k)^{n-1} + (-k)^n\right] + n^n.$$

Since n is odd, the terms k^n and $(-k)^n$ cancel each other out. $C_n^{n-1} n^1 (-k)^{n-1}$ is divisible by n^2 because $C_n^{n-1} = n$. All the remaining terms are divisible by n^2, because the exponents of n increase.

AL1.1.17

Starting from $k = 6$ and iterating the map "sum of squares of the digits" we obtain the chain 2006 – 40 – 16 – 37 – 58 – 89 – 145 – 42 – 20 – 4 – 16, after which the sequence repeats itself, with period 8. Thus, $f_1(2006) = 40$, $f_2(2006) = 16$, etc., and $f_{n+8}(2006) = f_n(2006)$ for all integers $n \geq 1$.

Since $2007 = 8 \cdot 250 + 7$, it follows that $f_{2007}(2006) = f_7(2006) = 42$.

AL1.1.18

The sum of the 43 differences d_i is $a_{44} - a_1 \leq 124$ (the maximum for a_{44} is 125 and the minimum for a_1 is 1).

If no value among the d_i's occurred more than 9 times, the sum would be at least $9 \cdot 1 + 9 \cdot 2 + 9 \cdot 3 + 9 \cdot 4 + 7 \cdot 5 = 125$, contradicting the above upper bound. Hence some value must occur at least 10 times.

AL1.1.19

Denoting $A = \lg a$, $B = \lg b$, $C = \lg c$, we have $A, B, C > 0$ (since $a, b, c > 1$) and the sum S becomes

$S = \dfrac{B}{A} + \dfrac{C}{A} + \dfrac{C}{B} + \dfrac{A}{B} + \dfrac{A}{C} + \dfrac{B}{C}$. By the arithmetic mean – geometric mean inequality, $S \geq 6\left(\dfrac{B \cdot C \cdot C \cdot A \cdot A \cdot B}{A \cdot A \cdot B \cdot B \cdot C \cdot C}\right)^{1/6} = 6$.

The example $A = B = C = 1$ shows that this bound is attained, so 6 is the smallest possible value of S.

AL1.1.20

Let n denote the number $\overline{abc}$, m the number $\overline{efg} = \overline{abc} - \overline{cba}$ and p the number $\overline{efg} + \overline{gfe}$. Then $n = 100a + 10b + c$ and $m = 100(a - c) - (a - c) = 100(a - c - 1) + 10 \cdot 9 + (10 - a + c)$, so that $e = a - c - 1, f = 9$, and $g = 10 - a + c$, due to the uniqueness of decimal representation. (The given conditions on a, b, c ensure that e, f, g fall in the interval $[0, 9]$.)

It follows that $p = 101e + 20f + 101g = 101 \cdot 9 + 20 \cdot 9 = 1089$.

AL1.1.21

Suppose not. Then there exist positive numbers a and b such that no pair of red points has distance a and no pair of black points has distance b. Without loss of generality we may assume $a \leq b$. Now consider a black point P; such a point has to exist, by our assumption that there are no two red points at distance a from each other (if all points were red, we could find two of them at any distance from each other, including a).

Consider now the circle of radius b around P. Since no two black points have distance b from each other, every point on this circle must be colored red. Since $a \leq b$, there exists two points on this circle having distance a from each other (the lengths of all chords take all the values of the interval $(0, 2b)$).

Thus, these two points are red points at distance a, which contradicts our assumption. Hence, one of the colors must contain, for every positive distance d, a pair of points at distance d.

AL1.1.22

Let $b_n = 1/a_n$. The given recurrence then takes the form $b_0 = 1$, $b_{n+1} = b_n + n$ $(n = 0, 1, 2, ...)$. Iterating this recurrence we obtain, for $n = 1, 2, ...$:

$$b_n = b_{n-1} + (n-1) = \cdots = b_0 + \sum_{k=0}^{n-1} k = 1 + \frac{1}{2}n(n-1) = \frac{1}{2}n^2 - \frac{1}{2}n + 1.$$

Hence, $a_n = \dfrac{1}{b_n} = \left(\dfrac{1}{2}n^2 - \dfrac{1}{2}n + 1\right)^{-1}$ $(n = 0, 1, 2, ...)$.

AL1.1.23

Set $g(x) = (x + 1)f(x) - x$. Then $g(x)$ is a polynomial of degree $n + 1$ which has roots at each of the $n + 1$ numbers $0, 1, \ldots, n$ (since $f(x)$ obeys the condition from hypothesis). Hence $g(x)$ must be of the form $g(x) = c\prod_{k=0}^{n}(x-k)$ for some constant c.

Setting $x = -1$ in the definition of $g(x)$ yields $g(-1) = 1$.
By the form of $g(x)$ it follows that

$$1 = g(-1) = c\prod_{k=0}^{n}(-1-k) = c(-1)^{n+1}(n+1)!.$$

Hence $c = (-1)^{n+1}/(n+1)!$. Substituting this value into the factorial expression of $g(x)$ and setting $x = n + 1$, we obtain

$$g(n+1) = c\prod_{k=0}^{n}(n+1-k) = \frac{(-1)^{n+1}}{(n+1)!}(n+1)! = (-1)^{n+1}, \text{ which implies}$$

$$f(n+1) = \frac{g(n+1) + n + 1}{n+2} = \frac{n+1+(-1)^{n+1}}{n+2}.$$

AL1.1.24

The given recurrence yields $x_{n+1} = x_n / y_n$, so that $y_n = x_n / x_{n+1}$ for all n. Hence $P_n = \prod_{k=1}^{n}(x_k / x_{k+1}) = x_1 / x_{n+1} = 1/x_{n+1}$ for all n.
Moreover, from the identity

$$y_n = \frac{1}{1+x_n} = \frac{1}{x_n} - \frac{1}{(1+x_n)x_n} = \frac{1}{x_n} - \frac{1}{x_{n+1}},$$ we see that

$$S_n = \sum_{k=1}^{n}\left(\frac{1}{x_k} - \frac{1}{x_{k+1}}\right) = \frac{1}{x_1} - \frac{1}{x_{n+1}} = 1 - \frac{1}{x_{n+1}}.$$

Hence $P_n + S_n = 1$ for all n.

AL1.1.25

Let $S_0 = 0$ and $S_n = a_1 + a_2 + \cdots + a_n$. By the counting principle (the pigeonhole principle), two of these $n + 1$ integers, say S_r and S_s (with $0 \leq r < s \leq n$), must leave the same remainder upon division by n. Hence $S_s - S_r = a_{r+1} + a_{r+2} + \cdots + a_s$ is a multiple of n.

AL1.1.26

Since $\log_2 3 > 0$, $\log_3 5 > 0$, $\log_5 8 > 0$, we can apply the arithmetic mean – geometric mean inequality and we obtain:
$$a > 3\sqrt[3]{\log_2 3 \cdot \log_3 5 \cdot \log_5 8} = 3\sqrt[3]{\log_2 8} = 3\sqrt[3]{3} = \sqrt[3]{81} > \sqrt[3]{64} = 4$$

Hence $a > 4$. (1)

We have:

$\log_2 3 < \log_2 4 = 2$ (2)

$3^3 > 5^2 \Rightarrow \log_3 5 < \dfrac{3}{2}$ (3)

$5^3 > 8^2 \Rightarrow \log_5 8 < \dfrac{3}{2}$ (4)

Adding together the inequalities (2), (3) and (4) yields $a < 5$. (5)

Finally, (1) and (5) yield $[a] = 4$.

AL1.1.27

Since an integer is divisible by 9 if and only if its sum of digits is divisible by 9, the numbers n with $f(n) = 9$ are exactly the multiples of 9. Since $2001 = 9 \cdot 222 + 3$, there are 222 such numbers below 2001.

AL1.1.28

Clearing the denominators and simplifying in the given relation yield $xyz + y^2z + yz^2 + x^2z + xyz + xz^2 + x^2y + xy^2 = 0$.

After grouping terms and factoring out, we obtain $(x+y)(x+z)(y+z) = 0$.

Hence $x = -y$, $x = -y$, or $y = -z$. In the first case, $x^n + y^n = 0$ for odd n, and so $x^n + y^n + z^n = (x+y+z)^n$. The other cases are analogous.

AL1.1.29

The answer is 99. To see this, evaluate N explicitly as follows:

$$N = (10-1) + (100-1) + \cdots + (\overbrace{10\ldots0}^{100}-1) = \overbrace{11\ldots10}^{99} - 99 = \overbrace{11\ldots10}^{97}11$$

AL1.1.30

We argue by contradiction. Suppose a, b and c are all odd and that $x = p/q$, with $(p, q) = 1$, is a rational solution of the given equation.

Clearing denominators yields $ap^2 + bpq + cq^2 = 0$. (1)

Since we assumed p and q are relatively prime, they cannot be both even. If p and q are both odd, then, in view of our initial assumption that a, b and c are odd, each term on the left of (1) is odd, so the left-hand side is odd and we have a contradiction. If exactly one of p and q is odd, then exactly two of the three terms on the left of (1) are even, an so the left-hand side is odd and we again arrive at a contradiction. Thus, a contradiction arises in either case, and a, b, and c therefore cannot all be odd.

AL1.1.31

Let $b_n = a_n - a_{n-1}$. The given recurrence for a_n yields $b_n = b_{n-2} + 1$ ($n \geq 3$), with initial conditions $b_1 = 1$, $b_2 = 1$. This implies, by induction, $b_{2n} = b_{2n-1} = n$ for all $n \geq 1$. Hence

$$a_{2n} = a_0 + \sum_{k=1}^{n}(a_{2k} - a_{2k-2}) = \sum_{k=1}^{n}(b_{2k} + b_{2k-1}) = \sum_{k=1}^{n}(2k) = n(n+1).$$

Hence $a_{2004} = 1002 \cdot 1003 = 1005006$.

AL1.1.32

If n is divisible by 4, then letting $a_1, a_2, \ldots, a_n$ be the pattern $(1, 1, -1, -1)$ repeated $n/4$ times, the terms $a_i a_{i+1}$ in the given sum is equal to 0. Thus, for all n divisible by 4, the given equation has a solution.

We now show that if n is not divisible by 4, there is no solution in integers ± 1. This is obvious in the case n is odd, since the left-hand side of the equation consists of a sum of an odd number of terms, each ± 1, and thus cannot be equal to 0.

It remains to consider the case when $n = 2m$, where m is odd. Suppose there exist integers $a_i = \pm 1$, $i = 1, 2, \ldots, n$, for which the given equation holds. Set $a_{n+1} = a_1$, so that the equation can be written as $\sum_{i=1}^{n} a_i a_{i+1} = 0$. Since $n = 2m$ and each of the terms $a_i a_{i+1}$ is ± 1, exactly m of these terms must be equal to 1, and m must be equal to -1. Hence the product of all $2m$ terms must be equal to $(1)^m (-1)^m = -1$, since m was assumed to be odd. On the other hand, a direct calculation shows that the product is equal to

$$\prod_{i=1}^{n} a_i a_{i+1} = \prod_{i=1}^{n} a_i^2 = 1,$$

so we have reached a contradiction. Thus, no solution exists when $n = 2m$ with m odd.

AL1.1.33

We apply Moivre's formula: $(\cos x + i \sin x)^n = \cos nx + i \sin nx$. For $x = 1/n$, this relation becomes

$$\left(\cos \frac{1}{n} + i \sin \frac{1}{n} \right)^n = \cos 1 + i \sin 1.$$

Unfolding the left-hand side, we obtain:

$$C_n^0 \cos^n \frac{1}{n} + C_n^1 \cos^{n-1} \frac{1}{n} i \sin \frac{1}{n} + \cdots + C_n^n i^n \sin^n \frac{1}{n} = \cos 1 + i \sin 1$$

The real terms from the left-hand side are

$$C_n^{2k} i^{2k} \cos^{n-2k} \frac{1}{n} \sin^{2k} \frac{1}{n}, \text{ with } k = 0, 1, \ldots, \left[\frac{n}{2}\right] \ (2\left[\frac{n}{2}\right] \text{ is the even}$$

number that is the nearest to n and less than or equal to n).

Hence $\sum_{k=0}^{\left[\frac{n}{2}\right]} C_n^{2k}(-1)^k \cos^{n-2k}\frac{1}{n}\sin^{2k}\frac{1}{n} = \cos 1$, which does not depend on n.

AL1.1.34

We have the inequality chain $[a]+[b] \leq [a+b] \leq [a]+[b]+1$ for any real numbers a and b. Making in particular $a = b$ yields $2[a] \leq [2a] \leq 2[a]+1$.

Since $2[a]$ and $2[a]+1$ are consecutive integers, the middle term in the inequality, which is also an integer, must equal one of these. We conclude that $2[a] = [2a]$ or $[2a] = 2[a]+1$, which is equivalent to $([2a]-2[a])([2a]-2[a]-1) = 0$ for any real number a.

Put $y = [2a]$, $x = [a]$, so that $P(x, y) = (y - 2x)(y - 2x - 1)$ satisfies the conditions of the problem.

AL1.1.35

We prove by complete induction. We can work out the first few numbers very easy. Assume that every number in $\{1, 2, \ldots, n-1\}$ can be properly represented. Consider n, $n > 3$.

If n is even, then $n/2 < n - 1$ is equivalent to $n > 2$, which is true. Hence $n/2$ can be properly represented. Multiplying every one of the terms in that representation by 2 (increasing the power of 2 by one in each case), we have a proper representation of n. If the given non-divisibility condition holds for $n/2$, then it also holds for n.

Now assume n is odd. There exists s such that $3^s \leq n < 3^{s+1}$. If $3^s = n$, we are done. If $3^s < n$, then let $m = \dfrac{n-3^s}{2}$. This is an integer because we assumed that n is odd. We have:

$$m = \frac{n-3^s}{2} \leq \frac{n-3}{2} < \frac{n-1}{2} < n-1$$

Then m can be properly represented. Furthermore, since

$$m = \frac{n-3^s}{2} < \frac{3^{s+1}-3^s}{2} = 3^s,$$ all the powers of 3 in the representation of m are smaller than 3^s. Multiplying that representation of m by 2 we get a representation of $2m$. Every one of the terms of this

representation of $2m$ contains a power of 2, and hence does not divide 3^s. All powers of 3 in the representation of $2m$ are smaller than 3^s, and hence are not divisible by 3^s.

Putting this representation of $2m$ together with 3^s we obtain a representation of n.

AL1.1.36
There are exactly the following possibilities:
$n = 1$, $k_i = 1$
$n = 3$, $k_i = 2, 3, 6$ (and permutations of that)
$n = 4$, $k_i = 4, 4, 4, 4$.

By arithmetic mean – harmonic mean inequality,

$$\frac{k_1 + \cdots + k_n}{n} \geq \left(\frac{1}{n}\left(\frac{1}{k_1} + \cdots + \frac{1}{k_n}\right)\right)^{-1}.$$ Using the given relations, this results in $\frac{5n-4}{n} \geq n$ or $n^2 - 5n + 4 \leq 0$. This factors to $(n-4)(n-1) \leq 0$. The only integer possibilities are $n = 1, 2, 3, 4$. Note also that the $n = 4$ case is the equality case of the AM – HM inequality, which forces all of the k_i's to be the same.

If $n = 1$, then $5n - 4 = 1$ and $k_1 = 1$, clearly the only possibility, and this case works.

If $n = 2$, then $5n - 4 = 6$. For the sum of the two reciprocals to be 1, neither of the k_i's can be 1, so each must be at least 2. But $\frac{1}{2} + \frac{1}{2} = 1$. No number larger than 2 would make this sum as big as 1. Since $2 + 2 \neq 6$, this case does not work and we have no solutions for $n = 2$.

If $n = 3$, then $5n - 4 = 11$. As before, each of the k_i's must be at least 2. If they are all at least 3, we must have them all equal to 3, so that $\frac{1}{3} + \frac{1}{3} + \frac{1}{3} = 1$. But $3 + 3 + 3 = 9 \neq 11$, so that does not work. Thus one of them must be 2. Now we are looking for integers b and c such that $\frac{1}{b} + \frac{1}{c} = \frac{1}{2}$ and $b + c = 9$. We cannot have $b = 2$. The only possibilities that make the fraction sum work are (3, 6) and (4, 4). With $3 + 6 = 9$ and $4 + 4 = 8$, only the former case works. So the k_i's must be 2, 3, 6.

Finally, if $n = 3$, then $5n - 4 = 16$. We noted that in this case the k_i's must be equal. The only possibility is 4, 4, 4, 4.

AL1.1.37

From Cauchy inequality we have
$$\left(\frac{1}{a^3(b+c)} + \frac{1}{b^3(c+a)} + \frac{1}{c^3(a+b)}\right)(a(b+c) + b(c+a) + c(a+b)) \geq \left(\frac{1}{a} + \frac{1}{b} + \frac{1}{c}\right)^2.$$

Since $abc = 1$, we have $\frac{1}{a} + \frac{1}{b} + \frac{1}{c} = ab + bc + ca$. Replacing in the above inequality, we obtain that
$$\left(\frac{1}{a^3(b+c)} + \frac{1}{b^3(c+a)} + \frac{1}{c^3(a+b)}\right) \geq \frac{(ab+bc+ca)^2}{2ab+2bc+2ca} = \frac{ab+bc+ca}{2}.$$

From arithmetic mean – geometric mean inequality, we have $ab + bc + ca \geq 3\sqrt[3]{a^2b^2c^2} = 3$. Therefore,
$$\left(\frac{1}{a^3(b+c)} + \frac{1}{b^3(c+a)} + \frac{1}{c^3(a+b)}\right) \geq \frac{3}{2}.$$

AL1.1.38

Let $x_1, x_2, \ldots$ be the given sequence and let $s_n = x_1 + x_2 + \cdots + x_n$, with the convention $s_0 = 0$.

The conditions from the hypothesis can be written as $s_{n+7} < s_n$ and $s_{n+11} > s_n$ for all $n \geq 1$. We then have:
$$0 < s_{11} < s_4 < s_{15} < s_8 < s_1 < s_{12} < s_5 < s_{16} < s_9 < s_2 < s_{13} < s_6 <$$
$$< s_{17} < s_{10} < s_3 < s_{14} < s_7 < 0,$$ which is a contradiction. Therefore, the sequence (s_n) cannot have 17 terms, so the sequence (x_n) cannot have 17 terms.

In order to show that 16 is the answer, we just take 16 real numbers satisfying

$s_{10} < s_3 < s_{14} < s_7 < 0 < s_{11} < s_4 < s_{15} < s_8 < s_1 < s_{12} < s_5 < s_{16} < s_9 < s_{13} < s_6$.

We have $x_1 = s_1$ and $x_n = s_n - s_{n-1}$ for $n \geq 2$. Thus we found all sequences with the given properties.

AL1.1.39
We have:

$$x^y = \underbrace{x \cdot x \cdots \cdot x}_{y} \cdot 1 \leq \left(\frac{yx+1}{1+y}\right)^{y+1} = \left(\frac{yx+x+1-x}{1+y}\right)^{y+1} = \left(x + \frac{1-x}{1+y}\right)^{y+1}.$$

We applied the arithmetic mean – geometric mean inequality.
The equality holds if and only if $x = 1$. Conversely, if $x = 1$, the given equation is satisfied.
Therefore the solution is any pair $(1, y)$, with y positive integer.

AL1.1.40
We claim that $x \geq y + 2$.
If $x = y$ the equation becomes $2x = 1$, which is impossible (x is natural number). If $x = y + 1$ the equation becomes $y + 1 = 2x + 1$, equivalent to $y + 1 = 2y + 3$, that is $y = -2$, which is impossible (y is positive). Therefore $x \geq y + 2$, as claimed.

We consider the following cases, by the values of y:
$y = 1$. The equation becomes $x = x + 1$, which has no solutions.
$y = 2$. The equation becomes $C_x^2 = x + 2 \Leftrightarrow x^2 - 3x - 4 = 0$, so $x = 4$.
$y \geq 3$. We claim that $C_x^y > x + y$ for every $x \geq y + 2$ and we argue by induction on x:
For $x = y + 2$ the inequality is equivalent to:
$C_{y+2}^y > 2y + 2 \Leftrightarrow C_{y+2}^2 > 2y + 2 \Leftrightarrow (y+1)(y+2) > 4y + 4$
$\Leftrightarrow y^2 - y - 2 > 0 \Leftrightarrow (y-2)(y+1) > 0$, which is true since $y \geq 3$.
Assume now $C_x^y > x + y$. We have $C_{x+1}^y = C_x^y + C_x^{y-1}$. Since $C_x^y > x + y$ and $C_x^{y-1} > 1$, adding them together results in $C_{x+1}^y = C_x^y + C_x^{y-1} > x + y + 1$ and the induction is complete. Hence in case $y \geq 3$ the equation has no solutions.
Therefore the only solution is $x = 4$, $y = 2$.

AL1.1.41

From the hypothesis we have $N = 100a + 10b + c = n^2$ and $100a + 10c + b = (n+1)^2$. Subtracting the first relation from the second yields $9(b - c) = 2n + 1$. It follows immediately that $b - c$ is odd, since the right-hand member is odd.

On the other hand, since the square of n has three digits, we have $10 \leq n+1 \leq 31$, which is equivalent to $19 \leq 2n+1 = 9(c-b) \leq 61$ or $3 \leq c-b \leq 6$. But $c - b$ is odd, therefore $c - b = 3$ or $c - b = 5$.

If $c - b = 3$, then $n = 13$, which satisfies the conditions of the problem. If $c - b = 5$, then $n = 22$, which does not satisfy those conditions. Hence the searched number is $n = 13$.

AL1.1.42

Suppose there exist natural numbers m and n such that $m^2 + 4n$ and $n^2 + 4m$ are perfect squares. Since the expressions are symmetrical in m and n, we can set an order between them, let us say $m \leq n$.

Since $n^2 + 4m > n^2$ and $n^2 + 4m$ is a square, it follows that $n^2 + 4m \geq (n+1)^2$.

On the other hand, $n^2 + 4m \leq n^2 + 4n < (n+2)^2$.

From the above inequalities necessarily results $n^2 + 4m = (n+1)^2$, which is equivalent to $4m = 2n + 1$, which is a contradiction, as the left-hand member is even and the right-hand member is odd.

AL1.1.43

a) In the given recurrence relation, we give values to n from 2 to n:

$x_2 = px_1 + qx_0; \ x_3 = px_2 + qx_1; \ \ldots \ x_n = px_{n-1} + qx_{n-2}$. Adding together these equalities yields $\sum_{i=2}^{n} x_i = p\sum_{i=1}^{n-1} x_i + q\sum_{i=0}^{n-2} x_i$, which becomes:

$$(p+q-1)\sum_{i=2}^{n-2} x_i = -qx_0 - (p+q)x_1 - x_{n-1}(p-1) + x_n$$

Since $p \geq 1, q \geq 0$ we have $p+q-1 \geq 0$. In addition, since $x_0, x_1 > 0$ we have $\sum_{i=2}^{n-2} x_i > 0$ and hence the right-hand member must be positive, which is equivalent to $x_n \geq x_{n-1}(p-1) + qx_0 + (p+q)x_1$.

b) We found that
$$(p+q-1)\sum_{i=2}^{n-2} x_i = -qx_0 - (p+q)x_1 - x_{n-1}(p-1) + x_n.$$
Adding $(p+q-1)(x_0 + x_1 + x_{n-1} + x_n)$ in both sides yields:
$$(p+q-1)\sum_{i=0}^{n} x_i = (p-1)x_0 - x_1 + qx_{n-1} + (p+q)x_n$$
Since $x_0 = 1$ and $x_1 = p-1$ we have $\sum_{i=0}^{n} x_i = \dfrac{(p+q)x_n + qx_{n-1}}{p+q-1}$.

Because p, q, x_0, x_1 are integers, it follows that $\sum_{i=0}^{n} x_i$ is also integer. Therefore expression E is divisible by $p + q - 1$.

AL1.1.44

If x would be larger than 1, multiplying N by 2, 3, 4, 5 or 6 would result at some point in a number having more than six digits. So x can only be 1.

No digit can be zero, as being the first digit of one of the given numbers. Due to the order relation between the six given numbers, we have $x < z < y < u < v < t$. (1)

We have:
$N = 10^5 x + 10^4 y + 10^3 z + 10^2 t + 10u + v$
$2N = 10^5 z + 10^4 t + 10^3 u + 10^2 v + 10x + y$
$3N = 10^5 y + 10^4 z + 10^3 t + 10^2 u + 10v + x$
$4N = 10^5 u + 10^4 v + 10^3 x + 10^2 y + 10z + t$
$5N = 10^5 v + 10^4 x + 10^3 y + 10^2 z + 10t + u$
$6N = 10^5 t + 10^4 u + 10^3 v + 10^2 x + 10y + z$
$2N \vdots 2 \Rightarrow y \in \{2, 4, 6, 8\}$ (2)
$3N \vdots 3 \Rightarrow x+y+z+t+u+v$ is multiple of 3 (3)
$4N \vdots 4 \Rightarrow 10z + t$ is multiple of 4 (4)

$5N \vdots 5 \Rightarrow u = 5$ (5)

$6N \vdots 2$ and $6N \vdots 3 \Rightarrow z \in \{2, 4, 6, 8\}$ (6)

Relations (1), (2) and (6) yield $x = 2, y = 4$.

Relations (1) and (4) imply $10z + t = 20 + t$ is multiple of 4, so $t = 8$.

Relations (1) and (3) imply $x + y + z + t + u + v$ is multiple of 3, so $20 + v$ is multiple of 3, and since $5 < v < 8$, we have $v = 7$.

Therefore: $N = 142857$; $2N = 258714$; $3N = 428571$; $4N = 571428$; $5N = 714285$; $6N = 857142$.

AL1.1.45

Let $a_1, a_2, \ldots, a_n$ be the integers that M consists of. We consider the subsequences $\{a_1\}, \{a_1, a_2\}, \ldots, \{a_1, a_2, a_3, \ldots, a_n\}$ and the sums $S_1 = a_1, S_2 = a_1 + a_2, \ldots, S_n = a_1 + a_2 + a_3 + \cdots + a_n$.

If one of numbers $S_1, S_2, \ldots, S_n$ is divisible by n (let this be $S_j = a_1 + \ldots + a_j$), the problem is solved and the searched subsequence is $\{a_1, \ldots, a_j\}$. In the opposite case, numbers $S_1, S_2, \ldots, S_n$ give remainders from the set $\{1, 2, \ldots, n-1\}$ upon division by n, so at most $n-1$ remainders. According to the counting principle, there exist two numbers S_i and S_j, $1 \le i < j \le n$, giving the same reminder upon division by n. It follows that $S_i - S_j$ is divisible by n. But $S_i - S_j = a_{i+1} + a_{i+2} + \cdots + a_j$ and hence the searched subsequence is $\{a_{i+1}, a_{i+2}, \ldots, a_j\}$.

AL1.1.46

Since $ab > 0$, we have two cases: $a > 0$ and $b > 0$, respectively $a < 0$ and $b < 0$. We prove the implication in the first case.

From $(a + b)c = 2ab$ follows $c = \dfrac{2ab}{a+b}$. Since $a \ge c$ and $b \ge c$ we have the inequalities $a \ge \dfrac{2ab}{a+b}$ and $b \ge \dfrac{2ab}{a+b}$.

Since $a > 0$ and $b > 0$, we are allowed to divide these inequalities by a and b respectively, yielding:

$a + b \ge 2b$ and $a + b \ge 2a \Leftrightarrow a \ge b$ and $b \ge a \Rightarrow a = b$

But c is the harmonic mean of numbers a and b and since these are equal, results $a = b = c$.

The case $a < 0$, $b < 0$ can be solved analogously, because it reverts to the first case if we denote $a = -x$ and $b = -y$, where $x, y > 0$.

AL1.1.47

Let $x_i, x_j \in E$ with $f(x_i) = x_j$. We have $f(f(x_i)) = x_i$, or $f(x_j) = x_i$. It follows that function f admits an inverse and $f^{-1} = f$.

Function f is bijective, therefore invertible.

If $x_i \in E$ (the domain of function f), then the image of x_i through f, namely $f(x_i)$, still belongs to set E (the range of function f).

Assume by absurdum that for any $x \in E$, $x \neq f(x)$.

Then the elements of E can be grouped in pairs, linked through the relations $x_j = f(x_i)$ and $x_i = f(x_j)$. Therefore, the number of elements of E is even, which contradicts the hypothesis.

Hence there exists a number $k \in E$ that is equal to its own image, that is $f(k) = k$.

AL1.1.48

Since the given relation holds for all n, we choose $n = 1$. We have $a_1^3 = a_1^2$ and, since a_1 is strictly positive, it follows that $a_1 = 1$.

We choose now $n = 2$ and we obtain $1 + a_2^3 = (1 + a_2)^2$, equivalent to $a_2^2 - a_2 - 2 = 0$, whose unique positive solution is $a_2 = 2$.

Assume $a_n = n$. We will prove by induction that $a_{n+1} = n+1$.

Replacing $n \to n+1$ in the given relation and using the induction hypothesis ($a_1 = 1, a_2 = 2, \ldots, a_n = n$), we obtain:

$$1^3 + 2^3 + \cdots + n^3 + a_{n+1}^3 = (1 + 2 + \cdots + n + a_{n+1})^2.$$

But $1 + 2 + \cdots + n = \dfrac{n(n+1)}{2}$ and $1^3 + 2^3 + \cdots + n^3 = \dfrac{n^2(n+1)^2}{2}$.

Replacing these two expressions in the above relation, we obtain $a_{n+1}^3 = n(n+1)a_{n+1} + a_{n+1}^2$. We can divide by $a_{n+1} \neq 0$ and we get $a_{n+1}^2 - a_{n+1} - n(n+1) = 0$.

This quadratic equation in a_{n+1} has the roots $-n$ and $n+1$, so only one positive root, namely $a_{n+1} = n+1$. Hence $a_n = n$ for any $n \in N^*$.

AL1.1.49

In the case where n is even, we have:
$$2^n + 3^n = 2^{2m} + 3^{2m} = 4^m + 9^m = (5-1)^m + (10-1)^m =$$
$5a + (-1)^m + 10b + (-1)^m = 5c + 2 \cdot (-1)^m$, where a, b, c are natural numbers.

It follows that the last digit of the given number belongs to the set $\{2, 3, 7, 8\}$. But there is no natural number such that the last digit of its square belongs to this set (the only possibilities are 1, 4, 9, 6, 5, 0). Hence the given number cannot be a perfect square, if n is even.

In the case where n is odd, we have:
$2^n + 3^n = 2^{2m+1} + 3^{2m+1} = (3-1)^{2m+1} + 3^{2m+1} = 3p - 1$, with p natural number.

Assume by absurdum that the given number is perfect square. Since it is not divisible by 3, it must be of form $(3q \pm 1)^2$, where q is a natural number.

But $(3q \pm 1)^2 = 3(3q^2 \pm 2q) + 1 = 3t + 1$, with t natural number.

The number cannot be simultaneously of forms $3p - 1$ and $3t + 1$. Indeed, $3p - 1 = 3t + 1$ is equivalent to $3(p - t) = 2$ and this is impossible, because 3 is not a divisor of 2.

We arrived at a contradiction. Hence the given number cannot be a perfect square, if n is odd.

AL1.1.50

Because $a^n + \dfrac{1}{a^n} = \dfrac{1}{a^{-n}} + a^{-n}$, it suffices to prove the affirmation for n natural. We argue by induction.

For $n = 1$ the sentence is obvious.

For $n = 2$, we have $a^2 + \dfrac{1}{a^2} = \left(a + \dfrac{1}{a}\right)^2 - 2$, which is integer.

For $n = 3$, we have $a^3 + \dfrac{1}{a^3} = \left(a + \dfrac{1}{a}\right)^3 - 3\left(a + \dfrac{1}{a}\right)$, which is integer.

Assume that $a^n + \dfrac{1}{a^n}$ is integer for 1, 2, 3, ..., n. We have:

$$a^{n+1} + \dfrac{1}{a^{n+1}} = \left(a^n + \dfrac{1}{a^n}\right)\left(a + \dfrac{1}{a}\right) - \left(\dfrac{1}{a^{n-1}} + a^{n-1}\right),$$ which is integer, due to the induction hypothesis.

AL1.1.51
First, we consider the particular case $x = y$. The equation becomes $2x^n = z^n$. Of course, $z > x$.

Since $2 \mid z^n$ it follows that $2 \mid z$, so z can be written as $z = 2k_1$, where k_1 is positive integer. Replacing back in the equation yields $2x^n = (2k_1)^n = 2^n k_1^n \Rightarrow x^n = 2^{n-1} k_1^n \Rightarrow 2 \mid x^n \Rightarrow 2 \mid x$.

Therefore, x can be also written as $x = 2q_1$, where q_1 is positive integer.

Replacing x and z back in the initial equation yields $2q_1^n = k_1^n$. The same reasoning applies to numbers (q_1, k_1) and we find the positive integers (q_2, k_2) such that $q_1 = 2q_2$ and $k_1 = 2k_2$.

Thus, we can build a decreasing sequence of positive even integers (k_m), such that $k_{m+1} = \dfrac{1}{2}k_m$. But when arriving at $k_m = 2$, we get $k_{m+1} = 1$, which is no longer even. This is a contradiction and the case $x = y$ is solved (the equation has no solutions).

Consider now the case $x \neq y$.

We assume this equation has a solution and show that each of x, y and z is larger than n.

Obviously, z is larger than each of x and y.
Without loss of generality, assume $y > x$.
$z \geq y + 1$, since z and y are integers. We have:

$$x^n = z^n - y^n = (z-y)(z^{n-1} + z^{n-2}y + \cdots + y^{n-1}) >$$
$$1(x^{n-1} + x^{n-2}x + \ldots + x^{n-1}) = \underbrace{x^{n+1} + \cdots + x^{n+1}}_{n} = nx^{n-1} \quad \text{(We used}$$

$z - y \geq 1$, $z > x$ and $y > x$)

Therefore, $x^n > nx^{n-1}$, so $x > n$, which contradicts with the hypothesis.

AL1.1.52

If p is the common difference, then $a_n = a_0 + np$ for $n \geq 0$.

The case $a_0 = 1$ is trivial, so suppose $a_0 > 1$.

The integers in the progression are those at least a_0 that are congruent to a_0 modulo p.

We are given positive integers a and b such that $a^j \equiv a_0 \equiv b^k \pmod{p}$ and $a^j, b^k \geq a_0 > 1$.

Since j and k are relatively prime, there exist integers t and u such that $jt + ku = 1$, and thus positive integers r and s such that $jr + ks \equiv 1 \pmod{(p-1)}$.

By Fermat's Little Theorem, for any such r and s,
$$\left(a^s b^r\right)^{jk} \equiv \left(a^j\right)^{ks}\left(b^k\right)^{jr} \equiv a_0^{ks} a_0^{jr} \equiv a_0^{ks+jr} \equiv a_0 \pmod{p}$$

All instances of $\left(a^s b^r\right)^{jk}$ are perfect jk-th powers in the progression.

AL1.1.53

With θ denoting the angle between the sides with lengths p and q, the area A is given by $A = \dfrac{1}{2} pq \sin\theta$. (1)

Also, $x^2 = p^2 + q^2 - 2pq \cos\theta$. (2)

Assume, without loss of generality, that $p \geq q$.

If $q = 2$, then (1) and (2) imply $x^2 = p^2 + 4 \pm 4\sqrt{p^2 - A^2}$.

Since the radical is rational, it must be an integer. Thus $x \equiv p \pmod{2}$.

The sides of the triangle satisfy $p - 2 < x < p + 2$.

Therefore, $x = p$, but then $p^2 - A^2 = 1$, which is impossible (because p is prime).

Hence, we may assume that p and q are odd.

Since (1) and (2) imply that $\sin\theta$ and $\cos\theta$ are rational, and (1) implies that the numerator of $\sin\theta$ is even, there exist positive integers a and b with $\gcd(a, b) = 1$ and $a \not\equiv b \pmod{2}$ such that

$$\sin\theta = \frac{2ab}{a^2 + b^2} \quad \text{and} \quad \cos\theta = \frac{a^2 - b^2}{a^2 + b^2} \quad (3)$$

Now (1) or (2) implies that $(a^2 + b^2) \mid pq$. Hence, (i) $a^2 + b^2 = pq$, (ii) $a^2 + b^2 = p$ or (iii) $a^2 + b^2 = q$.

The number of possible solutions (a, b) to any of these equals the number of possible values of x.

We claim that there are respectively at most four, one, and zero possible solutions, yielding at most five values of x.

It is well known that the number of distinct ordered pairs (a, b) satisfying (i) is four, if $p \equiv q \equiv 1 \pmod{4}$ and $p > q$. If $p = q \equiv 1 \pmod{4}$, then there are two solutions. Otherwise there are no solutions.

Now (ii), together with (3) and (2), implies that $x^2 = p^2 + q^2 - 2q(a^2 - b^2)$, from which it follows that $x \equiv \pm p \pmod{q}$ and x is even.

Since $p - q < x < p + q$, we cannot have $x \equiv p \pmod{q}$. For $x \equiv -p \pmod{q}$, there is exactly one value of x satisfying these constraints.

Finally, (iii), together with (3) and (2), implies that $x^2 = p^2 + q^2 - 2p(a^2 - b^2)$, from which it follows that $x \equiv \pm q \pmod{p}$. We still have that x is even and $p - q < x < p + q$, but this is impossible.

AL1.1.54

Suppose that (i) $a > b \geq 1$, (ii) $\gcd(a, b) = 1$, (iii) $(1 + ab) \mid (a^2 + b^2)$, and (iv) a is the smallest integer such that (i) through (iii) hold for some b. (This b cannot be 1).

There exist integers q and r such that $a = qb - r$ with $q \geq 2$ or $0 < r < b$. Note that $a \geq q$.

By (iii), $1 + ab$ divides $a^2 + b^2 - (q-1)(1+ab)$. From $a^2 = a(qb - r)$, we get
$$a^2 + b^2 - (q-1)(1+ab) = ab - ar + b^2 - q + 1 = a(b - r - 1) + b^2 + a - q + 1.$$

Since $r < b$ and $a \geq q$, this quantity is positive. It is bounded by $(a + b)b$, which is less than $2 + 2ab$ since $a > b$. By (iii), it is divisible by $1 + ab$, and hence it must equal $1 + ab$.

From $1 + ab = ab - ar + b^2 - q + 1$, we obtain $q = b^2 - ar$, and adding qbr to both sides yields $q(1 + br) = b^2 + (qb - a)r = b^2 + r^2$. Hence $(1 + br) \mid (b^2 + r^2)$ with $a > b > r \geq 1$. Since $b < a$, (iv) requires $r \mid b$. Since $(1 + br) \mid (b^2 + r^2)$ prohibits $r = 1$, now a and b have a non-trivial common factor r, contradiction.

AL1.1.55

Suppose that such set A exists. Since $y \notin A$ implies $y + m \in A$, we have $A \neq \phi$.

Consider x in A. Now $x + n \notin A$, so $x + n + m \in A$, $x + 2n + m \notin A$ and $x + 2n \in A$.

Repeating this argument yields $x + mn \in A$ if and only if m is even.

On the other hand, $x + n + m \in A$ implies $x + m \notin A$, so $x + 2m \in A$, starting from $x \in A$.

Repeating this yields $x + nm \in A$ if and only if n is even.

We thus have m even if and only if n is even, which contradicts the hypothesis that $m + n$ is odd.

AL1.1.56

If d_1 and d_2 are the lengths of the diagonals, Ptolomeu's inequality gives us: $a_1 a_3 + a_2 a_4 \geq d_1 d_2$.

Multiplying the two given inequalities yields
$$pq > \sqrt{(a_1^2 + a_2^2)(a_3^2 + a_4^2)}.$$

Then, applying Cauchy-Buniakowsky-Schwartz inequality, we obtain $pq > \sqrt{(a_1^2 + a_2^2)(a_3^2 + a_4^2)} \geq a_1 a_3 + a_2 a_4$.

But $a_1a_3 + a_2a_4 \geq d_1d_2$. It follows that $pq > d_1d_2$, therefore p and q cannot equal simultaneously to d_1 and d_2.

AL1.1.57

Let $S(n)$ denote the given sum. We claim that $S(n) = 1$ for all n. Since $S(1) = 1$, it suffices to show that $S(n + 1) = S(n)$ for all n.

Writing $k = n + 1 + h$ and using the identity $C_{n+1+h}^{n+1} = C_{n+1+h}^{h} = C_{n+h}^{h} + C_{n+h}^{h-1}$ for $h \geq 1$, we have:

$$2^{n+1} S(n+1) = \sum_{h=0}^{n+1} C_{n+1+h}^{h} 2^{-h} = \sum_{h=0}^{n+1} C_{n+h}^{h} 2^{-h} + \sum_{h=1}^{n+1} C_{n+h}^{h-1} 2^{-h}$$

$$= \sum_{h=0}^{n} C_{n+h}^{h} 2^{-h} + C_{2n+1}^{n+1} 2^{-n-1} + \sum_{h=0}^{n+1} C_{n+1+h}^{h} 2^{-h-1} - C_{2n+2}^{n+1} 2^{-n-2}$$

$$= 2^n S(n) + 2^n S(n+1) + \left(C_{2n+1}^{n+1} - \frac{1}{2} C_{2n+2}^{n+1} \right) 2^{-n-1}$$

Since $C_{2n+2}^{n+1} = C_{2n+1}^{n+1} + C_{2n+1}^{n} = 2C_{2n+1}^{n+1}$, the last term is zero, so we have $2^{n+1} S(n+1) = 2^n S(n) + 2^n S(n+1)$, and hence $S(n + 1) = S(n)$, as claimed.

AL1.1.58

We will show that the maximal cardinality sought is 171. To prove that the cardinality cannot exceed 171, suppose $A \subset \{1, 2, ..., 256\}$ is double-free. Given any element $a \in A$, let a_0 denote the odd part of a, so that $a = a_0 2^i$ with a_0 odd and i a non-negative integer. For each odd integer m, let A_m denote the set of $a \in A$ with $a_0 = m$. The sets A_m, $m = 1, 3, ..., 255$ partition the set A, so $|A| = |A_1| + |A_3| + \cdots + |A_{255}|$. To obtain an upper bound for $|A|$ we consider $|A_m|$ for different ranges of m.

If (1) $128 < m \leq 256$, then there can be at most one $a \in A$ with $a_0 = m$, namely $a = m$. Thus, the sum over $|A_m|$ for m in the range (1) is at most equal to the number of odd m in this range, i.e., 64.

If (2) $64 < m \leq 128$, then any $a \leq 256$ with $a_0 = m$ must be of the form $a = m$ or $a = 2m$, but because of the double-free condition at most one of these can belong to A. Hence $|A_m| \leq 1$ for m in the range (2), and the sum of $|A_m|$ over such m is at most 32.

If (3) $32 < m \leq 64$, then $a_0 = m$ implies that $a = m2^i$ with $i = 0, 1,$ or 2, but the double-free condition again implies that at most two of these can belong to A. Hence $|A_m| \leq 2$ in the range (3), and the sum of $|A_m|$ over m in this range is at most $16 \cdot 2 = 32$.

Similarly, considering the ranges (4) $16 < m \leq 32$, (5) $8 < m \leq 16$, (6) $4 < m \leq 8$, (7) $2 < m \leq 4$ (i.e., $m = 3$), and (8) $m = 1$, we see that $|A_m|$ is at most 2 in the range (4), 3 in the ranges (5) and (6), 4 in the range (7), and 5 in the range (8), and the corresponding sums over $|A_m|$ are bounded by $8 \cdot 2 = 16$, $4 \cdot 3 = 12$, $2 \cdot 3 = 6$, $1 \cdot 4 = 4$, and $1 \cdot 5 = 5$, respectively. Adding up these bounds, we obtain $|A| \leq 64 + 32 + 32 + 16 + 12 + 6 + 4 + 5 = 171$.

To show that this bound can be achieved, take A to be the set of integers $n \leq 256$ that are of the form $a = a_0 2^i$ with a_0 odd and $i = 0, 2, \ldots$. In this case, it is easy to check that the inequalities for $|A_m|$ in the above argument become equalities, and so we have $|A| = 171$.

AL1.1.59
The answer is $n = 166$. To see this, suppose first that $P(x)$ is a polynomial of degree n satisfying the three conditions (a), (b), and (c). Consider the polynomial $Q(x) = P(x) - x^{11}$. Then $Q(x)$ has degree at most n, and condition (a) implies that $Q(x)$ has a root at each of the numbers $k = 1, 2, \ldots, n$. It follows that $Q(x)$ is of the form $Q(x) = C(x-1)(x-2) \ldots (x-n)$ for some constant C. To determine C, we use the condition (c), which implies
$2003 = P(-1) = Q(-1) + (-1)^{11} = C(-1)^n(n+1)! - 1$.
Hence $C = 2004(-1)^n/(n+1)!$. Now,
$$P(0) = Q(0) = C(-1)^n n! = \frac{2004}{n+1},$$
so condition (b) holds if and only if $n+1$ is a divisor of 2004. Thus, any polynomial $P(x)$ of degree n satisfying all three conditions (a), (b), and (c) is of the form

$P(x) = C(x-1)(x-2) \ldots (x-n) + x^{11}$ with C as above and $n+1$ a divisor of 2004. Conversely, it is easy to see that any polynomial of this form satisfies (a), (b), and (c). Therefore the number n sought in the problem is the smallest divisor of 2004 that exceeds 13. Factoring 2004, we get $2004 = 2^2 \cdot 3 \cdot 167$. Hence, the smallest divisor exceeding 13 is 167, and so $n = 166$.

AL1.1.60

Let a set D and a sequence $(a_i)_{i=1}^{2006}$ with $a_i \in D$ be given as in the problem. For $n = 1, 2, \ldots, 2006$ let $P_n = \prod_{i=1}^{n} a_i$ denote the product of the first n terms and set $P_0 = 1$.

Note that any block of consecutive terms from the sequence $(a_i)_{i=1}^{2006}$ is of the form $\prod_{i=m+1}^{n} a_i = P_n / P_m$ for some integers m and n with $0 \le m < n \le 2006$. Thus, the problem amounts to showing that, for a suitable choice of (m, n) with $0 \le m < n \le 2006$, P_n / P_m is a perfect square. Since each a_i is among the numbers $d_1, d_2, \ldots, d_{10}$, each P_n is of the form $P_n = \prod_{i=1}^{10} d_i^{\alpha_{in}}$, where the exponents α_{in} are non-negative integers (with $\alpha_{i0} = 0$ for $i = 1, 2, \ldots, 10$; this also holds for P_0.)

Note that, by the definition of P_n as the product of the first n terms of the sequence (a_i), the exponents α_{in} are non-decreasing in n, for each i. Thus, for $0 \le m < n \le 2006$,

$$P_n / P_m = \prod_{i=m+1}^{n} a_i = \prod_{i=1}^{n} d_i^{\alpha_{in} - \alpha_{im}},$$

where the exponents $\alpha_{in} - \alpha_{im}$ are non-negative integers. Hence P_n / P_m will be a perfect square if the numbers $\alpha_{kn} - \alpha_{km}$, $k = 1, \ldots, 10$, are all even.

The latter condition holds if and only if the vectors $\alpha_n = (\alpha_{1n}, \ldots, \alpha_{10n})$ and $\alpha_m = (\alpha_{1m}, \ldots, \alpha_{10m})$ have the same parity in each component. Now, since each α_n, $n = 0, 1, 2, \ldots, 2006$, is a vector with 10 components, there are $2^{10} = 1024$ possible parity combinations for these components. Since we have 2007 < 1024, by the counting principle two of these must have the same parity combination. Denoting the indices of these two vectors by m and n (ordered so that $0 \le m < n \le 2006$), we then have that P_n / P_m is a perfect square, as claimed.

AL1.1.61

If $\sqrt{[x]} = \lfloor\sqrt{x}\rfloor$, then $\sqrt{[x]}$ is a non-negative integer, thus $[x] = p^2$ with p natural. Conversely, if $[x] = p^2$ with p natural, then $x = p^2 + \varepsilon$ with $\varepsilon \in [0, 1)$ and we have $\sqrt{[x]} = \sqrt{p^2} = p$.

Hence $\sqrt{[x]} = \lfloor\sqrt{p^2 + \varepsilon}\rfloor$. Since

$$p = \sqrt{p^2} \le \sqrt{p^2 + \varepsilon} < \sqrt{p^2 + 1} \le \sqrt{p^2 + 2p + 1} = \sqrt{(p+1)^2} = p+1,$$

we have $\lfloor\sqrt{p^2 + \varepsilon}\rfloor = p$. The solution of the given equation is the set $\bigcup_{p \in N} [p^2, p^2 + 1)$, or $[0, 1) \cup [1, 2) \cup [4, 5) \cup [9, 10) \cup \cdots$.

Generalization:

Solve the equation $\sqrt[n]{[x]} = \lfloor\sqrt[n]{x}\rfloor$ in real non-negative numbers, where n is a natural number, $n > 2$. Analogously to the previous particular case, we come to $x = p^n + \varepsilon$ with $\varepsilon \in [0, 1)$ and we have $\sqrt[n]{[x]} = \sqrt[n]{p^n} = p$ and $\lfloor\sqrt[n]{x}\rfloor = \lfloor\sqrt[n]{p^n + \varepsilon}\rfloor$. Since

$$p = \sqrt[n]{p^n} \le \sqrt[n]{p^n + \varepsilon} < \sqrt[n]{p^n + 1} \le \sqrt[n]{(p+1)^n} = p+1,$$

we have $\lfloor\sqrt[n]{p^n + \varepsilon}\rfloor = p$. The solution of the general equation is the set $\bigcup_{p \in N} [p^n, p^n + 1)$.

AL1.1.62

a) We prove that n is blocky if and only if $n > 1$ and $n \equiv \pm 1$ or $n \equiv \pm 7 \pmod{24}$. We want to find integers k, n, and m such that

$$m^2 = \frac{1}{n}\sum_{i=k}^{k+n-1} i^2 = \frac{1}{n}\left(\sum_{i=1}^{k+n-1} i^2 - \sum_{i=1}^{k-1} i^2\right).$$

Using the standard formula $\sum_{i=1}^{N} i^2 = N(N+1)(2N+1)/6$ and algebraically simplifying, we obtain
$m^2 = k^2 + (n-1)k + (n-1)(2n-1)/6$.

Thus $(n-1)(2n-1)/6$ must be an integer, which forces $n \equiv \pm 1$ (mod 6). Completing the square on the right side yields
$m^2 = (k + (n-1)/2)^2 + (n^2 - 1)/12$, so
$$\frac{n^2 - 1}{12} = m^2 - \left(k + \frac{n-1}{2}\right)^2 = \left(m + k + \frac{n-1}{2}\right)\left(m - k - \frac{n-1}{2}\right).$$
Since $n = 6l \pm 1$, $(n-1)/2 = 3l$ or $3l - 1$, while $(n^2 - 1)/12 = 3l^2 \pm l$, which is an even integer. Because $m + k + (n-1)/2$ and $m - k - (n-1)/2$ have the same parity and their product is even, we have $(n^2 - 1)/12 \equiv 0$ (mod 4). This forces $n \equiv \pm 1$ or $n \equiv \pm 7$ (mod 24). Conversely, if n is of this form, then $(n^2 - 1)/12 = (2\alpha)(2\beta)$, where α and β are integers. Taking $m = \alpha + \beta$ and $k = \alpha - \beta - (n-1)/2$ proves that the conditions are sufficient.

b) Given a blocky integer n, taking all possible factorizations of the form $(n^2 - 1)/12 = (2\alpha)(2\beta)$ with $\alpha, \beta > 0$ and setting $k = \alpha - \beta - (n-1)/2$ yields all possible starting points k for the summation. (Notice that if $\alpha = -a$ and $\beta = -b$ with $a, b > 0$, then k also arises from the positive factorization $(n^2 - 1)/12 = (2a)(2b)$, because $k = \alpha - \beta - (n-1)/2 = b - a - (n-1)/2$.)

c) From the construction above, we conclude that the number of such k is the number of positive divisors of $(n^2 - 1)/48$, which is easily computed from its prime factorization.

AL1.1.63
a) The necessity of the condition. We assume the graph of polynomial $P(x)$ admits the center of symmetry $C(a, b)$. This means:
$P(a - h) + P(a + h) = 2b$ for any real h (1)
We make the substitution $y = x - a$ and let $R(x)$ be a polynomial such that $P(x) = P(y + a) = R(y)$. (2)

For $x = a - h$ we obtain $y = a - h - a = -h$ and for $x = a + h$ we obtain $y = a + h - a = h$. Relation (1) becomes $R(-h) + R(h) = 2b$, for any real h.

Let $R(h) = a_0 + a_1 h + a_2 h^2 + \cdots + a_n h^n$, so we have
$$[a_0 - a_1 h + a_2 h^2 + \cdots + (-1)^n a_n h^n] + (a_0 + a_1 h + \cdots + a_n h^n) = 2b.$$
By identifying the coefficients, we obtain $a_0 = b$; $a_2 = a_4 = a_6 = \cdots = 0$.

Therefore $R(h) = b + a_1 h + a_3 h^3 + \cdots + a_{2k+1} h^{2k+1} = b + hQ(h^2)$, where $2k + 1$ is the largest odd number less than or equal to n, and $Q(h) = a_1 + a_3 h + \cdots + a_{2k+1} h^{2k+1}$. Because of relation (2), we have:
$$P(x) = R(y) = b + yQ(y^2) = b + (x-a)Q\left[(x-a)^2\right]$$

b) The sufficiency of the condition. Assume there exists a polynomial $Q(x)$ such that $P(x) = b + (x-a)Q\left[(x-a)^2\right]$. Let h be real. We take one at a time $x = a - h$ and $x = a + h$. We obtain $P(a-h) = b - hQ(h^2)$ and $P(a+h) = b + hQ(h^2)$, so $P(a-h) + P(a+h) = 2b$ for any real h. This means the graph of polynomial $P(x)$ admits point $C(a, b)$ as a center of symmetry.

AL1.1.64

$$f(x) = \frac{(x^2+1)+1}{\sqrt{x^2+1}} = \sqrt{x^2+1} + \frac{1}{\sqrt{x^2+1}}, \forall x \in R.$$

Let $a = \sqrt{x^2+1} > 0$. We consider the expression $g(a) = a + \frac{1}{a}$.

$(a-1)^2 \geq 0 \Leftrightarrow a^2 - 2a + 1 \geq 0 \Leftrightarrow a^2 + 1 \geq 2a$. Since $a > 0$ we are allowed to divide the last inequality by a, yielding $a + \frac{1}{a} \geq 2$, so $f(x) \geq 2, \forall x \in R$, that is $f(R) \subset [2, +\infty)$.

Conversely, let $u \in [2, +\infty)$. We consider the equation $f(x) = u$. From $u = \frac{x^2+2}{\sqrt{x^2+1}}$ it follows that $u\sqrt{x^2+1} = x^2+2$, and by raising to the second power we get $x^4 + (4-x^2)x^2 + 4 - u^2 = 0$.

After solving this biquadratic equation (making the substitution $x^2 = y$, etc.), we find the real solution $x = \sqrt{\dfrac{u^2 - 4 + u\sqrt{u^2 - 2}}{2}}$.

Hence for any $u \in [2, +\infty)$ there exists $x \in R$ such that $f(x) = u$, therefore $[2, +\infty) \subset R$.

From the two proven inclusions results $f(R) = [2, +\infty)$.

AL1.1.65

If set A has n elements then set $A \times A$ has n^2 elements and set $\mathcal{P}(A)$ has 2^n elements. For bijections between the two sets to exist, it is necessary that $n^2 = 2^n$.

We will solve this equation in natural numbers. Number $n = 1$ is not a solution, number $n = 2$ is a solution, number $n = 3$ is not a solution, while number $n = 4$ is a solution of the equation.

We claim that, if $n \geq 5$ then $2^n > n^2$. We argue by complete induction. For $n = 5$, obviously $2^5 > 5^2$. Assume $2^n > n^2$. We have: $2^{n+1} = 2 \cdot 2^n > 2n^2 = n^2 + n^2 > n^2 + 2n + 1 = (n+1)^2$. We used that $n^2 > 2n + 1$ for any $n \geq 5$. The proof of this last inequality is simple: $n^2 - 2n - 1 = n^2 - 2n + 1 - 2 = (n-1)^2 - 2$; we can observe that the right-hand member is strictly positive for any $n \geq 3$, so it is also for $n \geq 5$.

Hence $2^n > n^2$ for any $n \geq 5$, therefore the only solutions of the equation $n^2 = 2^n$ are $n = 2$ and $n = 4$.

For $n = 2$ we have 4! bijections and for $n = 4$ we have 16! bijections.

AL1.1.66

Let $y_1, y_2, \ldots, y_n$ n strictly positive numbers. Applying the arithmetic mean – geometric mean inequality, we have:

$$\frac{y_1}{y_2} + \frac{y_2}{y_3} + \cdots + \frac{y_n}{y_1} \geq n \sqrt[n]{\frac{y_1 y_2 \cdots y_n}{y_2 y_3 \cdots y_n y_1}} = n$$

Let a_i, $i = 1, \ldots, n$ be the coefficients of polynomial P.

We replace in the above inequality numbers y_i with each of $x_i^n, x_i^{n-1}, \ldots, x_i^2, x_i$, one at a time, then multiply the obtained inequalities by $a_0, a_1, \ldots, a_{n-1}$ respectively. We deduce:

$$a_0\left[\left(\frac{x_1}{x_2}\right)^n + \left(\frac{x_2}{x_3}\right)^n + \cdots + \left(\frac{x_n}{x_1}\right)^n\right] \geq a_0 n$$

$$a_1\left[\left(\frac{x_1}{x_2}\right)^{n-1} + \left(\frac{x_2}{x_3}\right)^{n-1} + \cdots + \left(\frac{x_n}{x_1}\right)^{n-1}\right] \geq a_1 n$$

$$\cdots\cdots\cdots\cdots\cdots\cdots\cdots\cdots\cdots\cdots\cdots\cdots$$

$$a_{n-1}\left(\frac{x_1}{x_2} + \frac{x_2}{x_3} + \cdots + \frac{x_n}{x_1}\right) \geq a_{n-1} n$$

In addition, we have $na_n \geq na_n$.

Adding together all $n + 1$ inequalities from above yields:

$$P\left(\frac{x_1}{x_2}\right) + P\left(\frac{x_2}{x_3}\right) + \cdots + P\left(\frac{x_n}{x_1}\right) \geq n(a_0 + a_1 + \cdots + a_n) = nP(1)$$

In the inequality Cauchy-Buniakovsky-Schwarz
$$(p_1^2 + p_2^2 + \cdots + p_n^2)(q_1^2 + q_2^2 + \cdots + q_n^2) \geq (p_1 q_1 + p_2 q_2 + \cdots + p_n q_n)^2$$
we set $p_1 = p_2 = \cdots = p_n = 1$ and
$$q_1 = P\left(\frac{x_1}{x_2}\right), q_2 = P\left(\frac{x_2}{x_3}\right), \ldots, q_n = P\left(\frac{x_n}{x_1}\right), \text{ yielding:}$$

$$n\left[P^2\left(\frac{x_1}{x_2}\right) + P^2\left(\frac{x_2}{x_3}\right) + \cdots + P^2\left(\frac{x_n}{x_1}\right)\right] \geq \left[P\left(\frac{x_1}{x_2}\right) + P\left(\frac{x_2}{x_3}\right) + \cdots + P\left(\frac{x_n}{x_1}\right)\right]^2$$

But $P\left(\frac{x_1}{x_2}\right) + P\left(\frac{x_2}{x_3}\right) + \cdots + P\left(\frac{x_n}{x_1}\right) \geq nP(1)$. Hence

$$P^2\left(\frac{x_1}{x_2}\right) + P^2\left(\frac{x_2}{x_3}\right) + \cdots + P^2\left(\frac{x_n}{x_1}\right) \geq \frac{n^2 P^2(1)}{n} = nP^2(1).$$

The equality holds if and only if $x_1 = x_2 = \cdots = x_n$.

AL1.1.67

Expression $E(a)$ does make sense for all a real (the expressions under the radicals are positive).

$E(a) = 0 \Leftrightarrow a = 0$

Because $E(-a) = \sqrt{a^2 - a + 1} - \sqrt{a^2 + a + 1} = -E(a)$, it suffices to determine the possible values of expression $E(a)$ for $a > 0$.

We have

$$E(a) = \frac{a^2 + a + 1 - (a^2 - a + 1)}{\sqrt{a^2 + a + 1} + \sqrt{a^2 - a + 1}} = \frac{2a}{\sqrt{a^2 + a + 1} + \sqrt{a^2 - a + 1}} > 0.$$

We claim that $E(a) < 1$, which is equivalent to

$2a < \sqrt{a^2 + a + 1} + \sqrt{a^2 - a + 1} \Leftrightarrow 4a^2 < 2(a^2 + 1) + 2\sqrt{a^4 + a^2 + 1} \Leftrightarrow$
$a^2 - 1 < \sqrt{a^4 + a^2 + 1} \Leftrightarrow a^2 - 1 < \sqrt{(a^2 - 1)^2 + 3a^2}$, which is obvious.

We prove now that for any $b \in (0, 1)$, there exist $a \in (0, +\infty)$ such that $E(a) = b$:

$\sqrt{a^2 + a + 1} - \sqrt{a^2 - a + 1} = b \Leftrightarrow 2(a^2 + 1) - 2\sqrt{a^4 + a^2 + 1} = b^2 \Leftrightarrow$
$2(a^2 + 1) - b^2 = 2\sqrt{a^4 + a^2 + 1} \Leftrightarrow 4(a^2 + 1)^2 - 4b^2(a^2 + 1) + b^4 =$
$= 4(a^4 + a^2 + 1) \Leftrightarrow 4a^2(1 - b^2) = 4b^2 - b^4 \Rightarrow a^2 = \frac{b^2(4 - b^2)}{1 - b^2} \Rightarrow$

$a = b\sqrt{\frac{4 - b^2}{1 - b^2}} > 0.$

Therefore, for $a \in (0, +\infty)$ we have $E(a) \in (0, 1)$. For $a \in (-\infty, 0)$, $E(a) \in (-1, 0)$ and hence $E(a) \in (-1, 1)$ for $a \in R$. So we proved that $E(R) = (-1, 1)$.

AL1.1.68

Denote $\sqrt{x} = a$, $\sqrt{y - 1} = b$, $\sqrt{z - 2} = c$. We have $x = a^2$, $y = b^2 + 1$, $z = c^2 + 2$. The given equation becomes $2a + 2b + 2c = a^2 + b^2 + c^2 + 3$, which is equivalent to $(a^2 - 2a + 1) + (b^2 - 2b + 1) + (c^2 - 2c + 1) = 0$, that is $(a - 1)^2 + (b - 1)^2 + (c - 1)^2 = 0$.

Since a, b, c are real numbers, it follows that $a = b = c = 1$ and therefore $x = 1, y = 2, z = 3$.

AL1.1.69

Since $\sqrt{7} - \dfrac{m}{n} > 0$, we have $7n^2 - m^2 > 0$. We search for the least $k \in N^*$ such that $7n^2 - m^2 = k$, that is $7n^2 = m^2 + k$. Therefore $m^2 + k$ must be a multiple of 7.

For $m = 7q$, we find $k = 7$.
For $m = 7q \pm 1$, we find $k = 6$.
For $m = 7q \pm 2$, we find $k = 3$.
For $m = 7q \pm 3$, we find $k = 5$.

These are all possible cases. Hence the least k is 3.
So we have $7n^2 - m^2 \geq 3$, which is equivalent to:

$$7 \geq \dfrac{m^2 + 3}{n^2} \Leftrightarrow \sqrt{7} \geq \dfrac{\sqrt{m^2 + 3}}{n} \Leftrightarrow \sqrt{7} - \dfrac{m}{n} \geq \dfrac{\sqrt{m^2 + 3} - m}{n}$$

It remains to show that $\dfrac{\sqrt{m^2 + 3} - m}{n} \geq \dfrac{1}{mn}$, which is equivalent to $\sqrt{m^2 + 3} \geq \dfrac{1}{m} + m \Leftrightarrow m^2 + 3 \geq \dfrac{1}{m^2} + 2 + m^2 \Leftrightarrow 1 \geq \dfrac{1}{m^2} \Leftrightarrow m \geq 1$, which is true.

AL1.1.70

We prove by induction. If $n = 2$, since f is bijective, we have: $f(a_1) = a_2$, $f(a_2) = a_1$ or $f(a_1) = a_1$, $f(a_2) = a_2$. In the first case we deduce $a_1 + a_2 < a_2 + a_1$, which is false. So the valid case is $f(a_k) = a_k$, $k = 1, 2$.

Assume the affirmation true for $n - 1$ numbers.

In case of n numbers, if $f(a_n) = a_n$ the affirmation reverts to the case with $n - 1$ numbers and is true. Assume (by absurdum) that $f(a_n) \neq a_n$. Since f is bijective, there exists $p < n$ such that $f(a_p) = a_n$.

We have $a_p + a_n < a_{p+1} + f(a_{p+1}) < \cdots < a_n + f(a_n)$. It follows that $a_p < f(a_{p+k})$, $k = 0, 1, 2, \ldots, n - p$ and hence there exist at least $n - p + 1$ numbers larger than a_p. But from the given order relation, the numbers larger than a_p are $a_{p+1}, a_{p+2}, \ldots, a_n$, namely $n - p$

numbers. We arrived at a contradiction, so we have $f(a_n) = a_n$ and thus $f(a_k) = a_k$ for $k = 1, \ldots, n$, which means that f is the identity function of set A.

If instead of A we have the set of integer numbers (which is infinite), we consider function $f : Z \to Z, f(x) = x + 1$.

The given inequalities are satisfied without f being the identity function of Z. So we found a counterexample, which proves that the affirmation is no longer true.

AL1.1.71

Assume the order $p_1 < p_2 < p_3$. From the given relation it follows that $q = \dfrac{p_1 + (p_2 \pm 1) + p_3}{3}$, which means that q is the arithmetic mean of numbers $p_1, p_2 \pm 1, p_3$ and therefore $p_1 < q < p_3$.

But $q \neq p_2$, because the opposite case would result in $2q = p_1 + p_3 \pm 1$, which is false, since in the left-hand member we have an even number and in the right-hand member we have an odd number, as p_1 and p_3 are odd.

But p_1, p_2, p_3 are consecutive primes and since $p_1 < q < p_3$ and $q \neq p_2$, we conclude that q cannot be prime, so it is composite.

AL1.1.72

Raising to the second power in the given relation yields
$a^2 a'^2 = b^2 c'^2 + b'^2 c^2 + 2bcb'c'$. (1)

Since the triangles are right-angled, by Pitagora's theorem, we have $a^2 a'^2 = (b^2 + c^2)(b'^2 + c'^2) = b^2 b'^2 + b^2 c'^2 + b'^2 c^2 + c^2 c'^2$.

Replacing this in relation (1), after we clear the identical terms, yields $b^2 b'^2 + c^2 c'^2 = 2bcb'c'$ or $(bb' - cc')^2 = 0 \Leftrightarrow bb' = cc'$ (2)

From the similarity of the two triangles follows the proportionality of their side-lengths, that is: $\dfrac{b}{b'} = \dfrac{c}{c'}$ (3)

Multiplying relations (2) and (3) we obtain $b^2 = c^2 \Rightarrow b = c$, which means that triangle of side-lengths a, b, c is isosceles.

Since the two triangles are similar, it follows that the second triangle is also isosceles.

AL1.1.73

If number x would have exactly one digit, then the highest value of expression $x + y + z$ is attained for $x = 9$. Hence we have $x + y + z \leq 27 < 60$, which contradicts with hypothesis.

If x would have exactly three digits, the lowest value of expression $x + y + z$ is attained for $x = 100$ and hence $x + y + z \geq 102 > 60$, which contradicts again with hypothesis. Obviously x cannot have more than three digits, as this minimum would become higher.

Therefore x has exactly two digits, hence it is of form $x = \overline{ab} = 10a + b$, with $a \in \{1, 2, \ldots, 9\}$ and $b \in \{0, 1, \ldots, 9\}$.

Then $y = a + b$. We have two cases to analyze:

a) $a + b \leq 9$. In this case $z = a + b$ and hence $x + y + z = 10a + b + (a + b) + (a + b) = 12a + 3b = 60$, equivalent to $b = 20 - 4a$.

We build the table of values for (a, b):

a	1	2	3	4	5	6	7	8	9
$b = 20 - 4a$	16	12	8	4	0	−4	−8	−12	−16

From this table it follows that the only possible values are (4, 4) and (5, 0), therefore $x = 44$ or $x = 50$.

b) $9 < a + b \leq 18$. In this case $a + b = \overline{1u} \Leftrightarrow a + b = 10 + u \Leftrightarrow 1 + u = a + b - 9 = z$. Hence $x + y + z = 10a + b + (a + b) + (a + b - 9) = 60$ and it follows that $12a + 3b = 69$, equivalent to $b = 23 - 4a$.

We build again the table of values for (a, b):

a	1	2	3	4	5	6	7	8	9
$b = 23 - 4a$	19	15	11	7	3	−1	−5	−9	−13

From this table it follows that the only possible values are (4, 7), therefore $x = 47$.

Thus we proved that the only numbers satisfying the problem's conditions are 44, 47 and 50.

AL1.1.74

From the given property it follows that $\left[\sqrt{n}\right] = \dfrac{n-1}{3}$, which implies $n - 1 = 3k$, that is $n = 3k + 1$, with $k \in N^*$.

We have $[3k+1] = \dfrac{3k+1-1}{3} = k$. By the definition of the integer part, this relation becomes:

$\sqrt{3k+1} - 1 < \left[\sqrt{3k+1}\right] \leq \sqrt{3k+1}$ or $\sqrt{3k+1} - 1 < k \leq \sqrt{3k+1}$.

The first inequality is equivalent to
$\sqrt{3k+1} < k+1 \Leftrightarrow k^2 + 2k + 1 > 3k + 1 \Leftrightarrow k > 1$. (1)

The second inequality is equivalent to
$k^2 \leq 3k + 1 \Leftrightarrow k^2 - 3k - 1 \leq 0$.

The equation $k^2 - 3k - 1 = 0$ has the roots $k_{1,2} = \dfrac{3 \pm \sqrt{13}}{2}$, so we have $\dfrac{3-\sqrt{13}}{2} \leq k \leq \dfrac{3+\sqrt{13}}{2}$.

Since $k \in N^*$ it follows that $1 \leq k \leq 3$. (2)

Relations (1) and (2) imply $k = 2$ or $k = 3$.

For $k = 2$ we have $n = 7$ and for $k = 3$ we have $n = 10$. These are the only natural numbers n with the given property.

AL1.1.75

Assume by absurdum that f is not injective. Then there exist $u, v \in N, u \neq v$ such that $f(u) = f(v)$, which is equivalent to:

$\{m^u \sqrt{2}\} = \{m^v \sqrt{2}\} \Leftrightarrow m^u \sqrt{2} - \left[m^u \sqrt{2}\right] = m^v \sqrt{2} - \left[m^v \sqrt{2}\right] \Leftrightarrow$

$\sqrt{2}(m^u - m^v) = \left[m^u \sqrt{2}\right] - \left[m^v \sqrt{2}\right] \Leftrightarrow \sqrt{2} = \dfrac{\left[m^u \sqrt{2}\right] - \left[m^v \sqrt{2}\right]}{m^u - m^v} \in Q$,

which contradicts with the fact that $\sqrt{2}$ is irrational.

AL1.1.76

a) $\sum\limits_{i=1}^{n} \lg x_i < 0 \Leftrightarrow \lg\left(\prod\limits_{i=1}^{n} x_i\right) < 0 \Leftrightarrow \left(\prod\limits_{i=1}^{n} x_i\right) \in (0, 1)$, from where it follows that at least one number x_k is smaller than 1 (in the opposite case, their product would be bigger than or equal to 1 and the logarithm of their product would be bigger than or equal to 0).

From the condition $\prod_{i=1}^{n} \lg x_i > 0$ it follows that among numbers $x_1, x_2, \ldots, x_n$ there exist an even number of numbers smaller than 1 (in the opposite case, the product of their logarithms would be negative or 0).

Hence, since there exists a number x_k smaller than 1, there exist at least two numbers smaller than 1 among numbers $x_1, x_2, \ldots, x_n$.

b) $\lg \left(\prod_{i=1}^{n} x_i \right) = p \Leftrightarrow \sum_{i=1}^{n} \lg x_i = p$.

We have
$s_n = \lg x_1 \lg x_2 + \lg x_1 \lg x_3 + \cdots + \lg x_{n-1} \lg x_n$, therefore:
$2s_n = 2\lg x_1 \lg x_2 + 2\lg x_1 \lg x_3 + \cdots + 2\lg x_{n-1} \lg x_n =$
$= \left(\sum_{i=1}^{n} \lg x_i \right)^2 - \sum_{i=1}^{n} \lg^2 x_i = p^2 - q$, hence $s_n = \frac{1}{2}(p^2 - q)$.

AL1.1.77

The numbers of form $\sqrt{m}$, with m prime, are irrational.
Indeed, for $m = 1$ the affirmation is true. For $m > 1$, if we assume by absurdum that $\sqrt{m} = \frac{a}{b}$, with a, b integers, and we take fraction a/b in lowest terms, it follows that $mb^2 = a^2$ and hence $m \mid a^2$ and, since m is prime, we deduce that $m \mid a$.

Therefore $a = mk$, with k natural. Replacing back, we obtain $mb^2 = m^2 k^2$, that is $b^2 = mk^2$.

By a similar argument, we obtain $m \mid b$. Hence $m > 1$ is a common divisor of numbers a and b, contradiction.

Denote these numbers $\sqrt{m}$ by $y_1, y_2, \ldots, y_m, \ldots$. We build the following infinite table:

$$\begin{array}{ccccc}
a_1x_1 & a_1x_2 & \ldots & a_1x_{n-1} & a_1x_n \\
a_2x_1 & a_2x_2 & \ldots & a_2x_{n-1} & a_2x_n \\
\ldots & \ldots & \ldots & \ldots & \ldots \\
a_mx_1 & a_mx_2 & \ldots & a_mx_{n-1} & a_mx_n \\
\ldots & \ldots & \ldots & \ldots & \ldots
\end{array}$$

We prove the required property by reductio ad absurdum. Assume that each row holds at least one rational number.

Since we have an infinity of rows and only n columns, according to counting principle it follows that there exists at least one column holding two rational numbers. Let this be the column with elements x_q. The two rational numbers would be of form a_ix_q and a_jx_q. Making their ratio, we get $\dfrac{a_ix_q}{a_jx_q} = \dfrac{a_i}{a_j}$.

But since a_i and a_j are irrational and the numbers under the radicals are prime, it follows that their ratio is irrational. We arrived at a contradiction, namely that a rational number is equal to an irrational one.

AL1.1.78

All terms with the rank higher than 3 are multiples of 10, because the product contains at least one even number and one number divisible by 5.

It follows that the remainder of N upon division by 10 is $1 + 2 \cdot 3 = 7$ and hence 7 is the last digit of N.

But there is no perfect square whose last digit is 7, because the perfect squares can only have 0, 1, 4, 5, 6 or 9 as their last digit.

AL1.1.79

We can write number 5^{7^n} in the following forms:
$5^{7^n} = 5^{(6+1)^n} = 5^{3k+1} = 125^k \cdot 5 = (31m+1)^k \cdot 5 = (31p+1) \cdot 5 = 31q + 5$.

Hence the remainder of the given number upon division by 31 is 5.

AL1.1.80
Let $f(x, y) = \sin 2x + 2\sin(x + y) - \sin 2y =$
$= 2\sin(x - y)\cos(x + y) + 2\sin(x + y) =$
$$= 2\sqrt{1+\sin^2(x+y)}\left(\frac{\sin(x-y)}{\sqrt{1+\sin^2(x+y)}}\cos(x+y) + \frac{1}{\sqrt{1+\sin^2(x+y)}}\sin(x+y)\right).$$

But $\left(\dfrac{\sin(x-y)}{\sqrt{1+\sin^2(x+y)}}\right)^2 + \left(\dfrac{1}{\sqrt{1+\sin^2(x+y)}}\right)^2 = 1$ and hence there exists $g(x, y) \in R$ such that:

$\dfrac{\sin(x-y)}{\sqrt{1+\sin^2(x+y)}} = \sin g(x, y)$ and $\dfrac{1}{\sqrt{1+\sin^2(x+y)}} = \cos g(x, y)$

Replacing them in the found expression of $f(x, y)$ yields:
$|f(x, y)| = 2\sqrt{1+\sin^2(x-y)}|\sin(x+y+g(x, y))| \le$
$\le 2\sqrt{1+\sin^2(x-y)} \le 2\sqrt{2}$.

AL1.1.81
First we show that among the 1000 numbers there is at least one odd number.

Indeed, in the opposite case all the 1000 numbers would be even and then their sum would be at least
$S = 2 + 4 + 6 + 8 + \cdots + 2000$. For calculating S, we group the terms in pairs, as follows: the first with the last, the second with the last but one, and so on. We find $S = 500 \cdot 2002 = 1001000$.

Because the sum of the 1000 numbers is lower than 1001000, it follows that there are also odd numbers among them. If only one odd number would exist among them, then the sum would be odd, contradiction. Hence at least two numbers must be odd.

AL1.1.82
If $x, y \in N$ and $x \ge 2, y \ge 2$ then numbers $1 + x!$ and $1 + y!$ are odd natural numbers and their product is also odd. On the other hand, $(x + y)!$ is even and the equation becomes impossible.

It remains to consider the following cases:
1) The case $x = 0$. In this case, the equation becomes $1 + y! = y!$, which has no solutions.
2) The case $x = 1$. In this case, the equation can be written as
$2(1 + y!) = (1 + y)! \Leftrightarrow 1 + y! = 3 \cdot 4 \cdot \ldots \cdot (y+1) \Leftrightarrow$
$1 = 3 \cdot 4 \cdot \ldots \cdot (y+1) - y! \Leftrightarrow 1 = 3 \cdot 4 \cdot \ldots \cdot y(y+1-2) \Leftrightarrow$
$2 = y!(y-1)$, having the only solution $y = 2$.

Due to the symmetry of the given equation, there is also the solution $x = 2$, $y = 1$.

Hence the solutions of the given equation are $(1, 2)$ and $(2, 1)$.

AL1.1.83
From the given equation and the fact that the exponential function takes only strictly positive values, it follows that $x > 0$.

Writing the arithmetic mean – geometric mean inequality for positive numbers $x \cdot 2^{\frac{1}{x}}$ and $\frac{1}{x} \cdot 2^x$ yields

$$x \cdot 2^{\frac{1}{x}} + \frac{1}{x} \cdot 2^x \geq 2\sqrt{2^{\frac{1}{x}+x}} = 2 \cdot 2^{\frac{\frac{1}{x}+x}{2}}.$$

We have $\frac{1}{x} + x \geq 2$. Indeed, this inequality is equivalent to $x^2 - 2x + 1 \geq 0 \Leftrightarrow (x-1)^2 \geq 0$, which is true for any real x.

Then we have $x \cdot 2^{\frac{1}{x}} + \frac{1}{x} \cdot 2^x \geq 4$ and the equality holds if and only if $x \cdot 2^{\frac{1}{x}} = \frac{1}{x} \cdot 2^x \Leftrightarrow x + \frac{1}{x} = 2 \Leftrightarrow x = 1$.

Hence the only solution is $x = 1$.

AL1.1.84
Because $225 = 9 \cdot 25$, it suffices to show that N is divisible by 9 and 25.

Because numbers $\underbrace{11\ldots10}_{n}$ and n are divisible by 5, it follows that their squares are divisible by 25, so the difference between these squares is also divisible by 25. It remains to prove the divisibility by 9.

We have:

$$N = \left(\underbrace{11\ldots10}_{n}-n\right)\left(\underbrace{11\ldots10}_{n}+n\right) =$$

$$\left[10^n+10^{n-1}+\cdots+10-\left(\underbrace{1+1+\cdots+1}_{n}\right)\right]\left(\underbrace{11\ldots10}_{n}+n\right) =$$

$$= \left[\left(10^n-1\right)+\left(10^{n-1}-1\right)+\cdots+\left(10-1\right)\right]\left(\underbrace{11\ldots10}_{n}+n\right) =$$

$$= \left(\underbrace{99\ldots9}_{n}+\cdots+99+9\right)\left(\underbrace{11\ldots10}_{n}+n\right) =$$

$$= 9\left(\underbrace{11\ldots1}_{n}+\cdots+11+1\right)\left(\underbrace{11\ldots10}_{n}+n\right)$$

This clearly shows that number N is divisible by 9.

AL1.1.85

We have $B - 2A = na^2 + nb^2 + pa^2 + pb^2 - 2na^2 - 2pb^2 =$
$= (p-n)a^2 - (p-n)b^2 = (p-n)(a^2-b^2) = (p-n)(a-b)(a+b)$,
from where it follows that $B - 2A$ is divisible by $(p-n)(a-b)$.

Since A is divisible by $(p-n)(a-b)$ and $B - 2A$ is divisible by $(p-n)(a-b)$, it follows that $B = (B - 2A) + 2A$ is divisible by $(p-n)(a-b)$.

AL1.1.86

Denote by i the height corresponding to side of lengths b. Then the area of the given triangle is $2n^2 = \dfrac{bi}{2} < \dfrac{bc}{2} < \dfrac{b^2}{2}$ (we used $i < c$ and $b > c$). Hence $4n^2 < b^2 \Leftrightarrow 2n < b$.

Similarly, we have $2n^2 = \dfrac{bi}{2} < \dfrac{ba}{2} < \dfrac{a^2}{2}$ (we used $i < a$ and $a > b$), from where it follows that $2n < a$. We add together the two obtained inequalities ($2n < b$ and $2n < a$): $4n < a + b$. We have $2n < 4n < a + b \Rightarrow a + b > 2n$.

Assume now the triangle is right-angled (the hypotenuse is necessarily a, as being the biggest side).

From $2n^2 < \dfrac{b^2}{2}$ and $2n^2 < \dfrac{a^2}{2}$ it follows that $n^2 < 2n^2 < \dfrac{b^2}{2}$ and $n^2 < 2n^2 < \dfrac{a^2}{2}$, which by addition gives $2n^2 < \dfrac{a^2 + b^2}{2}$. From Pitagora's theorem we have $b^2 = a^2 - c^2$. Thus, the last inequality becomes $4n^2 < 2a^2 - c^2$.

AL1.1.87

From $a_{n+1} = \sqrt{2}a_n - a_{n-1}, \forall n \geq 1$ it follows that
$a_{n+2} = \sqrt{2}a_{n+1} - a_n, \forall n \geq 1 \Leftrightarrow a_{n+2} = \sqrt{2}\left(\sqrt{2}a_n - a_{n-1}\right) - a_n =$
$= a_n - \sqrt{2}a_{n-1}, \forall n \in N^*$. Hence
$a_{n+4} = a_{n+2} - \sqrt{2}a_{n+1} = a_n - \sqrt{2}a_{n-1} - \sqrt{2}\left(\sqrt{2}a_n - a_{n-1}\right) = -a_n, \forall n \in N^*$.

Therefore $a_{n+8} = -a_{n+4} = a_n, \forall n \in N^*$, which means the sequence is periodic of unique period 8.

It follows that for all $n \in N^*$, $a_n \in \{a_1, a_2, \ldots, a_8\}$ and hence $a_n \in \{a_0, a_1, a_2, \ldots, a_8\}$ for all $n \in N$, from which we deduce that the sequence is bounded by $\min\limits_{1 \leq i \leq 8} a_i$ and $\max\limits_{1 \leq i \leq 8} a_i$.

AL1.1.88

We have $N_2 = N_1 - 1 = n^4 - 4n^3 + 8n^2 - 8n + 4 = \left(n^2 - 2n + 2\right)^2$, $N_1 = (n^2 - 2n + 2)^2 + 1$, so numbers N_1 and N_2 are consecutive and N_2 is a perfect square.

We claim that the number of divisors of the square of a natural number is odd. Let N be a perfect square and let d_1 be a divisor of N. Then there exists $d_2 \in N$ such that $N = d_1 d_2$. If $d_1 < \sqrt{N}$, then $d_2 > \sqrt{N}$. Indeed, assuming by absurdum that $d_2 \leq \sqrt{N}$ and multiplying the two inequalities, we obtain $N = d_1 d_2 < N$, contradiction.

Therefore, for any divisor smaller than $\sqrt{N}$ there exists a divisor bigger than $\sqrt{N}$. This means that the number of divisors of N is of form $2k + 1$ (we have added the divisor $\sqrt{N}$), so it is odd, as claimed.

Because for any x we have $x^2 < x^2 +1 < (x+1)^2$, it follows that $x^2 +1$ cannot be a perfect square. In particular, it follows that N_1 cannot be a perfect square, therefore p is even. Since q is odd (because N_2 is perfect square), it follows that $p + q$ is odd.

AL1.1.89

We can assume k is natural, because the representations of $-k$ in the required form can be obtained from the representations of k in this form, by switching each sign.

We argue the existence of a representation of natural number k in the required form by complete induction.

For $k = 0, 1, 2, 3$ there exist the representations:
$0 = 1^2 + 2^2 - 3^2 + 4^2 - 5^2 - 6^2 + 7^2$
$1 = 1^2$
$2 = -1^2 - 2^2 - 3^2 + 4^2$
$3 = -1^2 + 2^2$

We have the identity
$(m+1)^2 - (m+2)^2 - (m+3)^2 + (m+4)^2 = 4$, $\forall m \in N$. Indeed, this is equivalent to $2m - 4m - 6m + 8m - 12 + 16 = 4$, which is true for all m.

Assuming there exists a representation of k in form $k = \pm 1^2 \pm 2^2 \pm \cdots \pm m^2$, it follows that $k + 4$ can be also represented in this form, because
$k + 4 = \pm 1^2 \pm 2^2 \pm \cdots \pm m^2 + (m+1)^2 - (m+2)^2 - (m+3)^2 + (m+4)^2$.

According to the complete induction argument, it follows that any natural number k admits a representation in the required form.

If in the identity $(m+1)^2 - (m+2)^2 - (m+3)^2 + (m+4)^2 = 4$ we replace m with $m + 4$, we obtain:
$(m+5)^2 - (m+6)^2 - (m+7)^2 + (m+8)^2 = 4$, $\forall m \in N$.

Subtracting the two identities side by side yields:
$(m+1)^2 - (m+2)^2 - (m+3)^2 + (m+4)^2 - (m+5)^2 + (m+6)^2 +$

$+(m+7)^2 - (m+8)^2 = 0, \forall m \in N$

Thus, from the representation of k proved previously we can obtain another one, by replacing m with $m + 8$, and then another one, by replacing m with $m + 16$, and so on, obtaining an infinity of representations of k in the required form.

Hence any natural number k can be represented in an infinity of ways in form $k = \pm 1^2 \pm 2^2 \pm \cdots \pm m^2$, therefore any integer can be represented in an infinity of ways in this form, according to the first observation in this proof.

AL1.1.90

We have: $\log_b(n+1) > \dfrac{d}{bn} + \log_b n \Leftrightarrow \log_b\left(1+n^{-1}\right)^{nb} \geq b^d$.

We have $\left(1+\dfrac{1}{n}\right)^n \geq 2, \forall n \in N^*$. Indeed, for $n = 1$ the left-hand member becomes 2, so the equality holds. For $n = 2$, the left-hand member becomes $2,25 > 2$. Assume by absurdum that there exists a natural number $n > 2$ such that $\left(1+\dfrac{1}{n}\right)^n < 2$. Applying the logarithm of base n to this inequality, it becomes equivalent to $1+\dfrac{1}{n} < \log_m 2$. Since the left-hand member is positive and the right-hand member is negative, we arrived at a contradiction. Hence $\left(1+n^{-1}\right)^n \geq 2, \forall n \in N^*$. It follows that $\left(1+n^{-1}\right)^{nb} \geq 2^b > b^d$ and therefore the first inequality is true.

So we have $\log_b(k+1) > \dfrac{d}{bk} + \log_b k, \forall k \in N^*$ and summing over k from 1 to n yields:

$$\sum_{k=1}^{n} \log_b(k+1) > \frac{d}{b}\sum_{k=1}^{n}\frac{1}{k} + \sum_{k=1}^{n}\log_b k \Leftrightarrow$$

$$\log_b((n+1)!) > \frac{d}{b}\sum_{k=1}^{n}\frac{1}{k} + \log_b(n!) \quad (1)$$

We prove the second required inequality by induction. For

$n = 2$, the inequality becomes $\log_b 2 > \dfrac{d}{b} \cdot \dfrac{1}{2} \cdot 2 \Leftrightarrow 2^b > b^d$, which is true.

Assume $\log_b(n!) > \dfrac{dn}{b}\left(\dfrac{1}{2} + \dfrac{1}{3} + \cdots + \dfrac{1}{n}\right)$ (2)

and we prove that $\log_b((n+1)!) > \dfrac{d(n+1)}{b}\left(\dfrac{1}{2} + \dfrac{1}{3} + \cdots + \dfrac{1}{n} + \dfrac{1}{n+1}\right)$.

From (1) and (2) it follows:

$\log_b((n+1)!) > \dfrac{d}{b}\sum_{k=1}^{n}\dfrac{1}{k} + \dfrac{dn}{b}\sum_{k=2}^{n}\dfrac{1}{k} = \dfrac{d}{b}\left(\sum_{k=2}^{n}\dfrac{1}{k}\right)(n+1) + \dfrac{d}{b} =$

$= \dfrac{d(n+1)}{b}\sum_{k=2}^{n}\dfrac{1}{k} + \dfrac{d(n+1)}{b(n+1)} = \dfrac{d(n+1)}{b}\sum_{k=2}^{n+1}\dfrac{1}{k}$

According to the induction argument, it follows that

$\log_b(n!) > \dfrac{dn}{b}\left(\dfrac{1}{2} + \dfrac{1}{3} + \cdots + \dfrac{1}{n}\right), \forall n \geq 2$.

AL1.1.91

Each of the three factors in the left-hand member is necessarily a power of 3.

Let $x = 3^u$, $x + 2 = 3^v$, $x + 8 = 3^t$, such that $u + v + t = y$.

It follows that $3^v - 3^u = 2$ and $3^t - 3^u = 8$, which are equivalent to $3^u(3^{v-u} - 1) = 2$, respectively $3^u(3^{t-u} - 1) = 8$. It follows that $u = 0$ (2 and 8 do not have powers of 3 as divisors).

Hence $3^v - 1 = 2$, $3^t - 1 = 8$, that is $v = 1$ and $t = 2$.

Therefore, $x = 1$ and $y = 3$, which is the only solution of the equation.

AL1.1.92

We can write:

$a_{n+2} = \dfrac{2}{2 - a_{n+1}} = \dfrac{2}{2 - \dfrac{2}{2 - a_n}} = \dfrac{2 - a_n}{1 - a_n}$

We have $a_n \neq 1$, because $a_n = 1$ implies $a_{n+1} = \dfrac{2}{2-1} = 2$, which is a contradiction since $a_{n+2} = \dfrac{2}{2-a_{n+1}}$.

We can write:
$$a_{n+4} = \frac{2-a_{n+2}}{1-a_{n+2}} = \frac{2-\dfrac{2-a_n}{1-a_n}}{1-\dfrac{2-a_n}{1-a_n}} = \frac{-a_n}{-1} = a_n, \quad \forall n \in N$$

Hence the sequence is periodic of period 4.

AL1.1.93

Case 1. If $x > 0$, then $x^{2n} < 1 + x + \cdots + x^{2n} < (x+1)^{2n}$ (the first inequality is obvious and the second can be obtained directly from the Newton's development of the binomial in the right-hand member).

Hence we should have, according to the hypothesis, $x^{2n} < y^{2n} < (x+1)^{2n}$, which is equivalent to $x < y < x+1$ and this is impossible in integers (because x and $x+1$ are consecutive).

It follows that there is no integer y with the given property, hence the equation has no integer solutions in this case.

Case 2. If $x < -1$, we have similarly $(x+1)^{2n} < 1 + x + \cdots + x^{2n} < x^{2n}$. Indeed, we have:
$$1 + x + \cdots + x^{2n} = \frac{(1+x+\cdots+x^{2n})(1-x)}{1-x} = \frac{1-x^{2n+1}}{1-x} < x^{2n}$$
(the last inequality is equivalent to $1 < x^{2n}$, which is true).

For showing that $(x+1)^{2n} < 1 + x + \cdots + x^{2n}$, we use the denotation $1 + x = -a$, $a > 0$, so we have the equivalent inequality:
$$a^{2n} < \frac{(1+x+\cdots+x^{2n})(1-x)}{1-x} = \frac{1-x^{2n+1}}{1-x} = \frac{1+(1+a)^{2n+1}}{2+a}$$

The inequality $a^{2n} < \dfrac{1+(1+a)^{2n+1}}{2+a}$ is equivalent to $2a^{2n} + a^{2n+1} < 1 + (1+a)^{2n+1}$, which becomes obvious if we consider the coefficients of a^{2n} and a^{2n+1} in the development of the power of the binomial in the right-hand member.

Thus we proved that $(x+1)^{2n} < 1 + x + \cdots + x^{2n} < x^{2n}$ and it follows that $x + 1 < y < x$, which is impossible in integers.

So neither in this case the equation has any integer solutions.

Case 3. It remains to consider the situations $x = 0$, which implies $y = -1$ or $y = 1$, as well as $x = -1$, which implies $y = -1$ or $y = 1$. All these four pairs verify the given equation, therefore they are the only integer solutions.

AL1.1.94

We have $v_n = y_n - x_n$, $u_n = 2x_n - y_n$. Replacing back $u_{n+1}, v_{n+1}, u_n, v_n$ in the given relations yields $x_{n+1} = 3x_n + 2y_n$, $y_{n+1} = 4x_n + 3y_n$ and $x_0 = 5$, $y_0 = 7$.

From this recurrence it is obvious that y_n are positive integers.

We have $y_{n+1}^2 - 2x_{n+1}^2 = y_n^2 - 2x_n^2 = \cdots = y_0^2 - 2x_0^2 = -1$, so $y_n^2 - 2x_n^2 = -1$ for all n. From $2x_n^2 = y_n^2 + 1$ we deduce $\sqrt{2}x_n = \sqrt{1 + y_n^2} \Rightarrow [\sqrt{2}x_n] = [\sqrt{1 + y_n^2}]$. But $y_n^2 < y_n^2 + 1 < (y_n + 1)^2$ for $y_n > 0$, equivalent to $y_n < \sqrt{y_n^2 + 1} < y_n + 1$.

From this is follows that $[\sqrt{y_n^2 + 1}] = y_n$ and hence $[\sqrt{2}x_n] = y_n$.

AL1.1.95

One can easily check that polynomials $P = 0$ and $P = 1$ satisfy the given condition. Assume now $P \neq 0, P \neq 1$. Let $P \in R[X]$,
$P = a_n X^n + a_{n-1} X^{n-1} + \cdots + a_1 X + a_0$, $a_n \neq 0$.
$P(x_2) = P^2(x), \forall x \in R \Leftrightarrow a_n x^{2n} + a_{n-1} x^{2(n-1)} + \cdots + a_1 x^2 + a_0 =$
$= (a_n x^n + a_{n-1} x^{n-1} + \cdots + a_1 x + a_0)^2, \forall x \in R$

Assume that there exists $k \in \{0, 1, \ldots, n-1\}$ such that $a_k \neq 0$ and $a_s = 0$ for any $s \in \{k+1, k+2, \ldots, n-1\}$, which is equivalent to the fact that P is not of the form $a_n X^n$. Then we have:

$$P(x) = a_n x^n + a_k x^k + a_{k-1} x^{k-1} + \cdots + a_1 x + a_0$$

$$P(x^2) = a_n x^{2n} + a_k x^{2k} + a_{k-1} x^{2(k-1)} + \cdots + a_1 x^2 + a_0$$

The given relation becomes:

$$a_n x^{2n} + a_k x^{2k} + \cdots + a_1 x^2 + a_0 = \left(a_n x^n + a_k x^k + \cdots + a_1 x + a_0\right)^2, \forall x \in R$$

In the right-hand member of the above relation, the coefficient of x^{n+k} is $2a_n a_k \neq 0$, while in the left-hand member a term in x^{n+k} does not exist, since $2n > n + k > 2k$. Thus we arrived at a contradiction, so the assumption we made is absurd. Hence $P = a_n X^n$. Replacing back in the given relation yields:

$$P(x^2) = P^2(x), \forall x \in R \Leftrightarrow a_n x^{2n} = a_n^2 x^{2n}, \forall x \in R \Leftrightarrow a_n^2 = a_n \Rightarrow a_n = 1$$

Hence $P = X^n$. Therefore, the polynomials satisfying the given condition are $P \in \{0, 1, X^n\}$.

AL1.1.96

We show first that the possible remainders of a perfect square upon division by 7 are 0, 1, 2, or 4. Indeed, if $n = 7m + k$ with $k \in \{0, 1, 2, 3, 4, 5, 6\}$, then $n^2 = 49m^2 + 14mk + k^2$ and one can easily check that for $k = 0, 1, 2, 3, 4, 5, 6$ the remainders of k^2 upon division by 7 can be only 0, 1, 2, or 4. Hence the possible remainders of n^2 upon division by 7 are still 0, 1, 2, or 4. Thus we proved that every perfect square has this property.

Now let a, b, c, d, e be five integers. According to the above property, their squares can only have one of the remainders 0, 1, 2, or 4 upon division by 7. Since we have five numbers and only four remainders, according to the counting principle, there exist at least two squares that have the same remainder upon division by 7. Thus there exist $x, y \in \{a, b, c, d, e\}$ such that $x^2 - y^2 \vdots 7$. However

$x^2 - y^2 = (x - y)(x + y)$ and since 7 is prime number, it follows that 7 divides $x + y$ or $x - y$.

AL1.1.97

For fixed n, let $A = \{n, n + 1, \ldots, 2n\}$, $B = \{n! + n, n! + n + 1, \ldots, n! + 2n\}$ and let $f : A \to B$, $f(x) = x + n!$. Obviously, f is bijective (we can observe that $f(A) = B$, while the injectivity is obvious).

Let x be a composite number from set A, so x admits a non-trivial divisor k. Obviously, $k \le n$. Then $f(x) = x + n!$ is divisible by k, since each term is multiple of k. It follows that $f(x)$ is composite number. Therefore, through bijection f all composite numbers from A are "transported" in composite numbers from B, which shows that B has at least as many composite numbers as A has. It follows that B has at most as many prime numbers as A has.

However, the number of primes of B is $\pi(n!+2n) - \pi(n!+n)$ and the number of primes of A is $\pi(2n) - \pi(n)$. It follows that $\pi(n!+2n) - \pi(n!+n) \le \pi(2n) - \pi(n)$, which is equivalent to the claim.

AL1.1.98

For any $x \in R^*$ we have:

$$P_{n,m}\left(\frac{1}{x}\right) = \left(\frac{1}{x}-1\right)^n \left(\frac{1}{x}+1\right)^m = -\frac{P_{n,m}(x)}{x^{n+m}} \Leftrightarrow x^{n+m} P_{n,m}\left(\frac{1}{x}\right) = -P_{n,m}(x)$$

$$\Leftrightarrow x^{n+m} \sum_{i=0}^{n+m} a_i \frac{1}{x^i} = -\sum_{i=0}^{n+m} a_i x^i \Leftrightarrow \sum_{i=0}^{n+m} a_i x^{n+m-i} = -\sum_{i=0}^{n+m} a_i x^i,$$ from which,

through identification, we obtain $a_{n+m-j} = -a_j$, $\forall j \in \{0, 1, \ldots, n+m\}$.

Taking $j = \dfrac{n+m}{2}$ we get $a_{\frac{n+m}{2}} = -a_{\frac{n+m}{2}}$ and so $a_{\frac{n+m}{2}} = 0$.

Now, if for an index $j \in \left\{0, 1, 2, \ldots, \dfrac{n+m}{2}-1\right\}$ we have $a_j = 0$, it follows that for $k = n+m-j \in \left\{\dfrac{n+m}{2}+1, \dfrac{n+m}{2}+2, \ldots, n+m\right\}$ we also have $a_k = 0$.

Therefore, each null coefficient of index $j \in \left\{0, 1, 2, \ldots, \dfrac{n+m}{2}-1\right\}$ is in correspondence with a null coefficient of index $k \in \left\{\dfrac{n+m}{2}+1, \dfrac{n+m}{2}+2, \ldots, n+m\right\}$. Invoking $a_{\frac{n+m}{2}} = 0$, it follows that we have an odd number of null coefficients, that is $c_{n,m}$ is odd.

b) If $n = m$, then $P_{n,n}(x) = (x^2 - 1)^n$, which contains x only to even powers. Hence the coefficients of terms $x, x^3, x^5, \ldots, x^{2n-1}$ are null and these number n. Hence $c_{n,n} = n \geq 3$ and therefore the condition $n = m$ is sufficient for $c_{n,m} \geq 3$.

Further we show that this condition is not necessary, namely we show that there exists $P_{n,m}$ with $n \neq m$ such that $c_{n,m} \geq 3$.

Indeed, for $n = 7$ and $m = 15$ we have:
$$P_{7,15}(x) = (x-1)^7 (x+1)^{15} = (x^2 - 1)^7 (x+1)^8 =$$
$$= (-1 + C_7^1 x^2 - C_7^2 x^4 + \cdots)(1 + C_8^1 x + C_8^2 x^2 + C_8^3 x^3 + \cdots) =$$
$= a_0 + a_1 x + \cdots + a_{22} x^{22}$. We have $a_3 = -C_8^3 + C_7^1 C_8^1 = 0$ and according to the proof of point a) it follows that $a_{22-3} = a_{19} = 0$ and $a_{\underset{2}{22}} = a_{11} = 0$. Therefore $a_3 = a_{11} = a_{19} = 0$ and hence $c_{7,15} \geq 3$.

AL1.1.99

Let $k \in \{1, 2, \ldots, p-1\}$ be fixed. From the theorem of division we obtain $k^p = \mathcal{M} p^2 + a_k$, where $0 \leq a_k < p^2$. (1)

If $a_k = 0$ would hold true, then $k^p = \mathcal{M} p^2$, so p would divide k^p. Since p is prime, it follows that p would divide k, contradiction. Therefore, $0 < a_k < p^2$. (2)

Similarly we have $(p-k)^p = \mathcal{M} p^2 + a_{p-k}$, (3), where $0 < a_{p-k} < p^2$. (4)

We have:
$(p-k)^p = p^p - C_p^1 p^{p-1} k + \cdots + (-1)^{p-1} C_p^{p-1} p k^{p-1} + (-1)^p k^p =$
$= \mathcal{M} p^2 - k^p$, so $k^p + (p-k)^p = \mathcal{M} p^2$.

Adding together equalities (1) and (3) yields
$\mathcal{M} p^2 = \mathcal{M} p^2 + a_k + a_{k-p}$, so $a_k + a_{k-p} = \mathcal{M} p^2$. (5)

Adding together inequalities (2) and (4) yields
$0 < a_p + a_{p-k} < 2p^2$ and this together with (5) implies
$a_k + a_{k-p} = p^2$. (6)

Writing relation (6) for $k = 1, 2, \ldots, p-1$ and adding together the found equalities, we obtain $a_1 + a_2 + \cdots + a_{p-1} = p^2(p-1)$, which is exactly the claim.

AL1.1.100
We have:
$$P = n^4 + a = n^4 + 2n^2\sqrt{a} + a - 2n^2\sqrt{a} = \left(n^2 + \sqrt{a}\right)^2 - \left(n\sqrt{2\sqrt{a}}\right)^2 =$$
$$= \left(n^2 + \sqrt{a} + n\sqrt{2\sqrt{a}}\right)\left(n^2 + \sqrt{a} - n\sqrt{2\sqrt{a}}\right).$$

Let $\sqrt{2\sqrt{a}} = k \in N$. It follows that $2\sqrt{a} = k^2$ or $4a = k^4$, so $a = \dfrac{k^4}{4}$. For $k = 2m$ with $m \in N$ we get
$$P = \left(n^2 + 2m^2 + 2mn\right)\left(n^2 + 2m^2 - 2mn\right).$$

If $m > 1$ and hence $k > 2$, then $n^2 + 2m^2 + 2mn \geq 2m^2 \geq 8$ and $n^2 + 2m^2 - 2mn \geq (n-m)^2 + m^2 \geq m^2 \geq 4$. It follows that the two factors are bigger than 1, therefore P is not prime.

Thus, for $a \in \left\{8^2, 18^2, 32^2, \ldots, \left(2m^2\right)^2, \ldots\right\}$, P is not prime and so we found an infinity of such numbers.

AL1.1.101
If there exists a polynomial Q with the stated property, then $P(-x) = Q(-x)\,Q(x) = P(x)$, so polynomial function P is even.

Conversely, assume that polynomial function P is even. Let $P(x) = a_0 + a_1 x + \cdots + a_n x^n$, with $a_i \in C, \forall i = 1, \ldots, n$.

Writing $P(x) = P(-x)$, $\forall x \in C$ and identifying the coefficients, it follows that all coefficients of the terms with odd powers of x are null. Hence we can write:
$$P(x) = a_0 + a_2 x^2 + a_4 x^4 + \cdots + a_{2n} x^{2n} = P_1(x^2), \forall x \in C, \quad (1)$$
where P_1 is a polynomial of degree n.

If $x_1, x_2, \ldots, x_n \in C$ are the roots of polynomial P_1, we can write $P_1(x) = a(x - x_1)(x - x_2)\cdots(x - x_n)$ with $a \in C^*$ and then (1) becomes $P(x) = a(x^2 - x_1)(x^2 - x_2)\cdots(x^2 - x_n) =$

$$= (-1)^n a(x_1 - x^2)(x_2 - x^2) \cdots (x_n - x^2). \quad (2)$$

Let $y_1, y_2, \ldots, y_n \in C$ be complex numbers such that $y_i^2 = x_i$ $(i = 1, \ldots, n)$ and let $b \in C$ be a complex number such that $b^2 = (-1)^n a$. Relation (2) becomes:
$$P(x) = b^2(y_1^2 - x^2)(y_2^2 - x^2) \cdots (y_n^2 - x^2) =$$
$$= [b(y_1 + x)(y_2 + x) \cdots (y_n + x)] \cdot [b(y_1 - x)(y_2 - x) \cdots (y_n - x)].$$
Taking $Q(x) = b(y_1 + x)(y_2 + x) \cdots (y_n + x)$, we have $P(x) = Q(x)Q(-x), \; \forall x \in C$.

References

American Mathematical Monthly, Vol. 116, Number 10, December 2007, USA

American Mathematical Monthly, Vol. 115, Number 6, June-July 2008, USA

Kvant, Number 3, 1970, URSS

Kvant, Number 11, 1971, URSS

Kvant, Number 2, 1978, URSS

Gazeta Matematică, Number 4, 1977, Romania

Gazeta Matematică, Number 5, 1977, Romania

Gazeta Matematică, Number 12, 1977, Romania

Gazeta Matematică, Number 3, 1979, Romania

Gazeta Matematică, Number 8, 1979, Romania

Gazeta Matematică, Number 1, 1980, Romania

Gazeta Matematică, Number 8, 1980, Romania

Gazeta Matematică, Number 11, 1980, Romania

Gazeta Matematică, Number 4, 1981, Romania

Gazeta Matematică, Number 5, 1981, Romania

Gazeta Matematică, Number 9, 1981, Romania

Gazeta Matematică, Number 10, 1981, Romania

N. Teodorescu, A. Constantinescu, M. Mihai, L. Pârşan, E. Perjariu, A. Popescu-Zorica, P. Radovici-Mărculescu, M. Ţena, *Problems from Gazeta Matematică – Selective and Methodological Edition (Probleme din Gazeta Matematică – Ediţie selectivă şi metodologică)*. Editura tehnică, Bucharest, 1984.

INFAROM Publishing Project

This book is part of Infarom's project of publishing a series of math problem books with an interactive structure, compiled from collective contributions from mathematics instructors and graduates.
The content submissions are made on the basis of a collaboration that provide the contributors with a full author or co-author status, in various working and financial formulas. All the details about this project and the registration are posted on publisher's website at www.infarom.com/problem_books.html .

*From the applied mathematics collection,
by same author*:

"**Understanding and Calculating the Odds:
Probability Theory Basics and
Calculus Guide for Beginners, with
Applications in Games of Chance
and Everyday Life**"
ISBN 9738752019

"**Roulette Odds and Profits:
The Mathematics of Complex Bets**"
ISBN 9738752078

The books can be ordered from all major online bookstores.
For the orders made through publisher's website, a 30% discount applies:
www.infarom.com/new_releases.html

www.ingramcontent.com/pod-product-compliance
Lightning Source LLC
Chambersburg PA
CBHW072159160426
43197CB00012B/2448